C0 DVZ 583

# NEW FRONTIERS IN ANGIOGENESIS

# New Frontiers in Angiogenesis

*Edited by*

REZA FOROUGH
*Texas A&M University,*
*Health Science Center*
*College Station, TX, U.S.A.*

A C.I.P. Catalogue record for this book is available from the Library of Congress.

ISBN-10 1-4020-4326-0 (HB)
ISBN-13 978-1-4020-4326-0 (HB)
ISBN-10 1-4020-4327-9 (e-book)
ISBN-13 978-1-4020-4327-7 (e-book)

---

Published by Springer,
P.O. Box 17, 3300 AA Dordrecht, The Netherlands.

*www.springer.com*

*Printed on acid-free paper*



Printed in the Netherlands.

# TABLE OF CONTENTS

# PREFACE

Understanding cellular and molecular mechanisms involved in regulation of angiogenesis has been an area of intensive biomedical research. In broad terms, this interest stems from the fact that a comprehensive knowledge of how to control accentuation of "good" and attenuation of "bad" angiogenesis offers promising therapeutic modalities for ischemic cardiovascular diseases and solid cancers, respectively. Besides these two most common diseases of mankind, which are responsible for the highest human mortality rates, other common debilitating diseases including rheumatoid arthritis and diabetes-related disorders are angiogenesis-dependent.

Interestingly, the more we learn about angiogenesis and its contribution to health and disease states, the more we realize the urgent need to redefine this advancing field. For example, it is still widely assumed that the occurrence of vasculogenesis, the formation of new blood vessels in situ, is solely limited to the embryonic state. However, accumulating evidence, mainly due to advances in the stem cell research, support the concept that vasculogenesis can be initiated by circulating progenitor endothelial cells in adults. In other words, the current definition of vasculogenesis as embryonic formation and differentiation of the vascular system is no longer concise.

In the past few years, our knowledge about cellular and molecular mechanisms regulating angiogenesis has greatly expanded. More importantly, this new information will revolutionize our concept of angiogenesis, if it has not already done so. Among these revolutionary findings are that gene regulation of a vascular bed formed within a tumor mass (tumor angiogenesis) is significantly different from normal nearby vessels or the findings on close parallels between angiogenesis and neurogenesis which will bring the researchers in the fields in close collaborations as well as help emergence of new strategies for developing therapies to treat cancers and neuropathies of different sorts, respectively.

The kind of advances described above is the impetus for putting this book together.

*New Frontiers in Angiogenesis* is a fresh and unconventional look at the field of angiogenesis. Each chapter will take the reader along on a journey into uncharted territories of angiogenesis. This volume starts with a comprehensive overview of the field and continues with topics that have been minimally explored. The topics deal with dynamics of vasculogenesis using imaging techniques, bone marrow-derived endothelial cell precursors as potential therapeutic tools, regulation of post-angiogenic vessel regression, vascular mimicry, design and construction of artificial vessels, bioengineering of angiogenesis, and lymphangiogenesis recapitulating angiogenesis in health and disease states. Each chapter is written by leading experts of the subjects. It is hoped that this volume will challenge all of us interested in the field of angiogenesis to think "outside the box" and explore angiogenesis from a fresh angle. It is hoped that *New Frontiers in Angiogenesis* is thought provoking and serve as a road map for discovering new findings to help betterments of human health.

I would like to express my appreciations to all contributors for their excellent contributions and the Springer Publisher staff for their quality work and timely publication of this book. I would also like to thank Stephanie Hilliard and Chen Hu for their excellent assistance in putting this volume together. I also extend my appreciations to my lab members including Brian Weylie, Charles Collins, Gregory Chen, Elena Kozhina, and Chen Hu for stimulating discussions. The last but not least, I wish to thank my wife Arzu for her continuing support and encouragement, and to my sons, Shayan and Darian, who patiently put up with many hours they have to wait before they could get to play with their dad.

Finally, it is my hope that this volume serves as a platform to stimulate new ideas leading to new discoveries and the expansion of new frontiers in the field of angiogenesis.

Reza Forough
College Station, TX

# LIST OF CONTRIBUTORS

Christopher R. Anderson
University of Virginia, Department of Biomedical Engineering, Charlottesville, Virginia, USA

Alfred C. Aplin
Division of Pathology and Laboratory Medicine, VA Puget Sound Health Care System, Seattle, Washington, USA

Ola Awad
Department of Anatomy University of Iowa, Iowa City, Iowa, USA

John C. Chappell
University of Virginia, Department of Biomedical Engineering, Charlottesville, Virginia, USA

Martine Dunnwald
Department of Dermatology, University of Iowa, Iowa City, Iowa, USA

Reza Forough
Department of Medical Physiology & Cardiovascular Research Institute, The Texas A&M University System Health Science Center, College Station, Texas, USA

Mary J. C. Hendrix
Children's Memorial Research Center, Northwestern University Feinberg School of Medicine and Robert H. Lurie Comprehensive Cancer Center, Chicago, Illinois, USA

Charles D. Little
Department of Anatomy and Cell Biology, University of Kansas Medical Center, Kansas City, Missouri, USA

Jeanette A. M. Maier
Department of Preclinical Sciences, University of Milan, Milan, Italy

Massimo Mariotti
Department of Preclinical Sciences, University of Milan, Milan, Italy

Meghan M. Nickerson
University of Virginia, Department of Biomedical Engineering, Charlottesville, Virginia, USA

Beatrice Nico
Department of Human Anatomy and Histology, University of Bari Medical School, Bari, Italy

Roberto F. Nicosia
Division of Pathology and Laboratory Medicine, VA Puget Sound Health Care System, Seattle, WA; Department of Pathology, University of Washington, Seattle, Washington, USA

Richard J. Price
University of Virginia, Department of Biomedical Engineering, Charlottesville, Virginia, USA

Domenico Ribatti
Department of Human Anatomy and Histology, University of Bari Medical School, Bari, Italy

Stanley G. Rockson
Stanford Center for Lymphatic and Venous Disorders, Division of Cardiovascular Medicine, Stanford University School of Medicine, Stanford, California, USA

Paul A. Rupp
Stowers Institute for Medical Research, Kansas City, Missouri, USA

Gina C. Schatteman
Departments of Exercise Science, University of Iowa, Iowa City, Iowa, USA

Elisabeth A. Seftor
Children's Memorial Research Center, Northwestern University Feinberg School of Medicine and Robert H. Lurie Comprehensive Cancer Center, Chicago, Illinois, USA

Richard E. B. Seftor
Children's Memorial Research Center, Northwestern University Feinberg School of Medicine and Robert H. Lurie Comprehensive Cancer Center, Chicago, Illinois, USA

William S. Shin
Stanford Center for Lymphatic and Venous Disorders, Division of Cardiovascular Medicine, Stanford University School of Medicine, Stanford, California, USA

Ji Song
University of Virginia, Department of Biomedical Engineering, Charlottesville, Virginia, USA

Elisabetta Weber
Department of Neurosciences, Section of Molecular Medicine, University of Siena Medical School, Siena, Italy

Evan A. Zamir
Department of Anatomy and Cell Biology, University of Kansas Medical Center, Kansas City, Missouri, USA

Wen-Hui Zhu
Division of Pathology and Laboratory Medicine, VA Puget Sound Health Care System, Seattle, Washington, USA

Chapter 1

# ANGIOGENESIS: AN OVERVIEW

Massimo Mariotti and Jeanette AM Maier
*Department of Preclinical Sciences, University of Milan*

Tissue homeostasis is dependent upon an adequate supply of oxygen and nutrients delivered through blood vessels. Since the cells must be located within 100-200 μm of blood vessels, the diffusion limit for oxygen, to survive and properly function, the maintenance of the vascular network is critical for survival of tissues. Angiogenesis, derived from the Greek word angêion meaning vase, and genesis meaning birth, is the name given to the outgrowth of new capillaries from the pre-existing primary plexus and is crucial in responding to tissue demands both in physiological and pathological conditions.

The term angiogenesis was coined in 1794 by the British surgeon John Hunter to describe blood vessels growth in reindeer antlers as a result of long lasting exposure to cold[1], a situation that can be interpreted today as a response to vasoconstriction, and therefore increased luminal shear stress[2]. In 1935, the term angiogenesis was applied to medicine by the pathologist Arthur Hertig to describe the formation of new vessels in the placenta. Angiogenesis became a common word in the biomedical community after surgeon Judah Folkman proposed that tumor growth is dependent upon neovascularization[3]. Currently, dysregulated angiogenesis is considered a common denominator in most frequent diseases, including cancer, ischemic heart disease, blindness, psoriasis, and arthritis[4]. Consequently, understanding how blood vessels form has become a principal, yet challenging, objective of studies over the last decade.

The protagonist in angiogenesis is the endothelial cell and co-stars are mural cells as well as extracellular matrix (ECM) components. The endothelium is a dynamic, highly heterogeneous, disseminated organ that possesses vital secretory, synthetic, metabolic and immunologic functions[5]. Endothelial cells are among the longest-lived cells outside of the nervous



*R. Forough (ed.), New Frontiers in Angiogenesis,* 1–29.

system. In fact, physiological endothelial cell turnover is reportedly measured in years in tissues that do not require angiogenesis. In a normal adult vessel, only 1 in every 10000 endothelial cells (0.01%) is in the cell division cycle at any given time[6]. Mural cells, i.e. the smooth muscle cell in arterioles and arteries, or pericytes in the capillaries, are associated with the abluminal surface of the vessel. Pericytes are required for normal microvascular structure and function because they provide structural support, protect endothelial cells from apoptosis, and actively control the stability of the vessels[7,8]. Smooth muscle cells endow blood vessels with vasomotor properties, which are necessary to accommodate the changing needs in tissue perfusion.

During embryonic development, blood vessels develop from endothelial precursors called angioblasts which assemble into a primary capillary plexus. This process is referred to as vasculogenesis. The primitive network is then remodelled by angiogenesis, so that new blood vessels sprout and branch from pre-existing capillaries in order to complete the circulatory system[9]. Later in development, endothelial heterogeneity arises due to microenvironmental signals to the endothelial cells. The last step is the recruitment and organization of mesenchymal cells into layers, generating mature, functional blood vessels.

In the adult, new vessels are produced mainly through angiogenesis. The few adult tissues that require ongoing angiogenesis are the female reproductive organs. Angiogenesis occurs during the monthly reproductive cycle to rebuild the uterus lining and to mature the egg during ovulation, and during pregnancy to build the placenta, allowing circulation between mother and fetus. Angiogenesis is also necessary for the repair or regeneration of tissues during wound healing. These normal processes lead to the formation of a stable, operational vascular bed through a tightly regulated balance between pro- and anti-angiogenic signals. An alteration of this equilibrium promotes dysregulated vessel growth with a consequent major impact on health[10].

## 1. WHAT TRIGGERS ANGIOGENESIS

Under normal circumstances, angiogenesis is a highly ordered and regulated process driving quiescent endothelial cells into a series of events culminating with the organization of a vascular network that responds to the demands of the growing or healing tissues. Metabolic stress has a role in influencing angiogenesis since various signals from the microenvironment including low oxygen tension, low extracellular pH, and low glucose concentration, trigger neovascularization. The first experimental evidence that angiogenesis is subject to some form of metabolic regulation was provided by the striking relation in the muscle between capillary density and the metabolic rate[11]. More recently, it became clear that angiogenesis is triggered by low oxygen levels itself rather than by hypoxia-dependent consequences on energy metabolism[12]. Indeed, an enzyme, the prolyl 4-hydroxylase, binds molecular oxygen and links it to the hypoxia-inducible factor (HIF)-1α. Another protein, VHL, which stands for von Hippel Lindau, associates with oxydized HIF-1α promoting its ubiquitination and degradation through the proteasome[13]. When the oxygen concentration falls, this process slows and HIF-1α levels increase. HIF-1α is a key transcriptional regulator of several molecules involved in angiogenesis including Vascular Endothelial Growth Factor (VEGF), Platelet-derived Growth Factor (PDGF), and angiopoietin-2 [14-16]. Recently, the deletion of HIF-1α in endothelial cells was shown to inhibit endothelial proliferation, chemotaxis, extracellular matrix penetration, and wound healing. Most strikingly, loss of HIF-1α results in a profound inhibition of blood vessel growth in solid tumors. These phenomena are all linked to a decreased level of VEGF expression and loss of autocrine response of VEGFR-2 in HIF-1α null endothelial cells[17].

Also, mechanical forces mediated by blood flow have profound effects on vessel growth. In vivo and in vitro it has been shown that shear stress induced by blood flow modulates vessel morphogenesis. Vessels that are not perfused eventually regress, mainly by endothelial apoptosis[18]. The contribution of biomechanical forces to vascular remodelling is highlighted in the female reproductive system where high rates of blood flow associate with extensive endothelial proliferation[19].

Important contributors to angiogenesis are immune or inflammatory cells which infiltrate the tissues in pathological situations. Indeed, most leukocyte subtypes produce pro-angiogenic molecules[20]. Tumor associated macrophages, for instance, are an excellent source of angiogenic factors. Similarly, mast cells promote angiogenesis after encountering allergens and pathogens[21]. However, leukocytes also generate angiogenesis inhibitors and their role in initiating or terminating angiogenesis is dependent upon the temporal and spatial balance of these modulators. The role of immune and inflammatory cells is further complicated by the evidence that: i) leukocytes and endothelial cells influence each other; ii) some angiogenic molecules also modulate leukocyte functions[22]; and iii) some chemokines both recruit leukocytes and stimulate endothelial cells[23]. Due to the significant involvement of leukocytes in some diseases, the suppression of angiogenesis by anti-inflammatory drugs is expected[24].

Another trigger of angiogenesis, fundamental in tumors, is genetic mutation. Interestingly, the activation of oncogenes src snd ras or the deletion of tumor suppressor genes PTEN, p53, or VHL induces or amplifies the abovementioned HIF system, thus activating the angiogenic process[25].

## 2. THE MOLECULES INVOLVED IN ANGIOGENESIS

The molecular signalling pathways involved in activating endothelial cells to initiate angiogenesis are relatively well known. Extracellular signals are mainly secreted paracrine factors and ECM components. The list of the molecules involved in angiogenesis is very long and continuously growing (summarized in Figure 1-1).

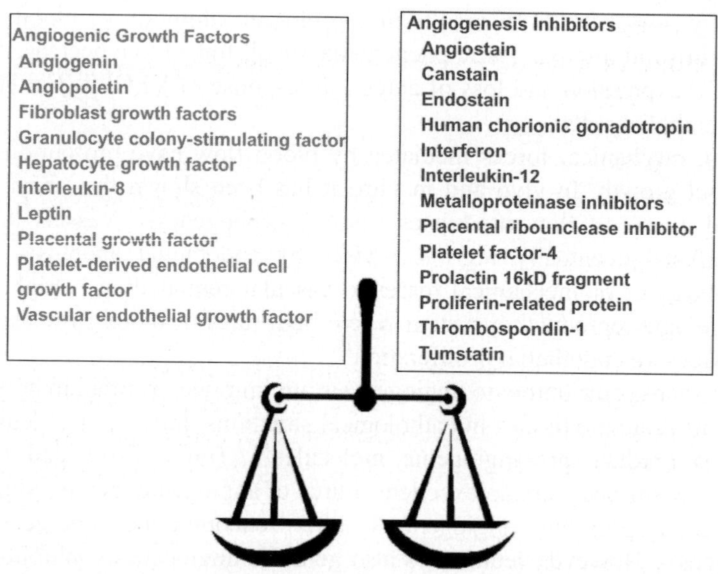


*Figure 1-1.* Angiogenesis results from the balance between pro- and anti-angiogenic factors.

Some factors are endothelial specific, such as the members of the VEGF and the angiopoietin families, and Ephrin B2 and 4B. Many other growth factors modulate the function of different cell types including endothelial cells, such as members of the Fibroblast Growth Factor (FGF), PDGF or Tranforming Growth Factor β (TGFβ) families. In addition, a myriad of other gene products - from Notch to transcription factors - have been shown crucial for vessel formation. Intriguingly enough, many genes thought to be specific to neurons also play a role in angiogenesis, suggesting a developmental similarity between nervous tissue and the vasculature[26]. Insoluble matrix components also control blood vessel growth (Table 1-1). All of these molecules must function in perfect harmony, in a complementary and coordinated manner to form functional vessels.

In such a redundant system, VEGF maintains its position as the most critical driver of vascular formation, as it is necessary in both vasculogenesis and angiogenesis during early development as well as in the adult. Alternative splicing of a single gene generates six isoforms of VEGF which, with a few exceptions (see below), demonstrate identical biological activities. The main receptors involved in initiating signal transduction cascades in response to VEGF comprise a family of closely related, endothelial specific receptor tyrosine kinases termed VEGFR-1, VEGFR-2 and VEGFR-3. VEGFR-3 is also required for lymphangiogenesis[27]. Accessory receptors, such as neuropilins, have also been reported[28]. VEGFR-2 appears to mediate most of the effects of VEGF, while VEGFR-1 may act either as a decoy receptor or by suppressing signalling through VEGFR-2[29]. VEGF is a protagonist in angiogenesis in all its phases since it increases vascular permeability, stimulates ECM remodelling, induces endothelial proliferation and migration, inhibits apoptosis, enhances branching of the neoformed vessels, and regulates lumen diameter. Indeed, disruption of even a single VEGF allele in mice leads to embryonic death[30]. VEGF inactivation in older animals is much less traumatic, affecting only those structures that continue to undergo vascular remodelling, such as corpus luteus[31]. As expected, unregulated VEGF expression in the absence of the entire angiogenic program, leads to the formation of immature and leaky vessels which can have devastating consequences[32]. VEGF alone is unable to direct blood vessel organization and maturation and must work in concert with other factors. The angiopoietins (Ang) seem to be among VEGF's most important partners. These proteins bind the Ties, a family of receptor tyrosine kinases selectively expressed on the vascular endothelium. Ang1 stimulates branching and remodelling of the newly formed vessels through its ability to increase the girth and stability of the endothelium. In the end, it stabilizes the vessel and maintains endothelial quiescence. Ang1 maximizes interactions between endothelial cells and the surrounding cells and matrix, thus counterbalancing the effects of VEGF on permeability. Ang2 antagonizes Ang1, thus destabilizing the vasculature in a manner that is necessary for its subsequent remodeling.

*Table 1-1.* Effects of some extracellular matrix proteins on angiogenesis.

| PRO-ANGIOGENIC PROTEINS | | |
|---|---|---|
| | Effects | Phenotype of null mice |
| Fibronection | promotion of endothelial adhesion, growth, survival; modulation of FGF and VEGF | defective vascular development<br>decreased angiogenesis after wounding |
| Vitronectin | promotion of endothelial survival | decreased angiogenesis after wounding |
| Perlecan | enhancement of FGF signalling | decreased vessel stability |
| Tenascein C | promotion of endothelial migration and sprouting | not available |
| Collagen I | promotion of tube formation | blood vessel rupture |
| Collagen III | stablization of vessel wall | blood vessel rupture |
| ANTI-ANGIOGENIC PROTEINS | | |
| Decorin | inhibition of endothelial migration and organization in tubes | not available |
| Thrombospondin-1 | inhibition of endothelial growth and survial | increased vascularity in the skin and pancreatic islets |
| Thrombospondin-2 | inhibition of endothelial growth | increased vascularity in various tissues |

Other factors may play a role in angiogenesis though they are not endothelial specific. Among others, the FGF family, which consists of more than twenty distinct but related members, has been considered for years to be involved in the formation of new vessels. Indeed, acidic and basic FGF, the most extensively studied members of the family, induce processes in endothelial cells in vitro that are critical to angiogenesis, since they stimulate protease synthesis, migration, proliferation and formation of a tube-like structure in three dimensional matrices[33]. Nevertheless, FGFs do not appear to play a crucial role in angiogenesis as vascular development is found to be normal in FGF-2 deficient mice[34].

The participation of PDGF is pivotal in maintaining capillary wall stability through the recruitment of pericytes to the neoformed vessels. In parallel, TGFβs cooperate in forming and strengthening the vessel wall by inhibiting endothelial cell proliferation as well as promoting mural cell differentiation.

Also membrane-bound factors should be taken into account. Integrins, cadherins and ephrins are endothelial membrane proteins that play prominent roles in angiogenesis. Endothelial cell adhesion molecules of the integrin family have emerged as critical mediators and regulators of angiogenesis. Indeed, integrins provide the physical interaction with the ECM necessary for cell adhesion, migration and positioning, and induction of signalling events essential for cell survival, proliferation and differentiation[35]. In particular, the $\alpha_v\beta_3$ and $\alpha_v\beta_5$ integrins have long been considered to positively regulate the angiogenic switch, because their pharmacological antagonists suppress angiogenesis in many experimental models and are currently being tested in clinical trials for their therapeutic efficacy against angiogenesis-dependent diseases. However, recent data using gene targeted mice revealed that the picture is far more complex. Surprisingly, increased tumor angiogenesis was observed in mice lacking the $\beta_3$ integrin[36], suggesting that $\beta_3$ integrins can also function as negative regulators of angiogenesis.

Vascular endothelial (VE)-cadherin, an endothelial cell-specific cadherin localized exclusively at the intercellular adherens junctions, regulates the passage of molecules across the endothelium, mediates contact inhibition of endothelial cell growth and enhances endothelial survival by promoting transmission of VEGF‘s antiapoptotic signal to the nucleus.

A growing body of evidence supports the importance of the concerted actions of ephrins and their receptors (Eph) in angiogenesis. Genetic studies, using targeted mutagenesis in mice reveal that ephrin-B1, ephrin-B2, and EphB4 are essential for the normal morphogenesis of the embryonic vasculature into a sophisticated network of arteries, veins, and capillaries[37]. In addition, engagement of Eph receptors by ephrin ligands mediates critical steps of angiogenesis, including juxtacrine cell-cell contacts, cell adhesion to the extracellular matrix, and cell migration. Recent evidence from in vitro angiogenesis assays and analysis of mice deficient for one or more members of the Eph family establish the role of Eph signalling in sprouting angiogenesis and blood vessel remodelling during vascular development. Furthermore, elevated expression of Eph receptors and ephrin ligands is associated tumor vasculature, suggesting that Eph receptors and their ephrin ligands also play critical roles in tumor angiogenesis[38].

The discovery of endogenous angiogenesis inhibitors underscores the finely tuned, sophisticated regulation of neovascularization since these negatively-acting molecules work in concert with angiogenic molecules to curtail the process. Inhibitors push the equilibrium in favour of vessel quiescence by acting at several levels: i) by inhibiting ECM remodelling through the inhibition of proteases; ii) by inhibiting endothelial proliferation, adhesion and migration; iii) by promoting endothelial apoptosis.

Thrombospondin-1, a large matricellular glycoprotein secreted by many cell types, was one of the first endogenous inhibitors of angiogenesis to be discovered. This large multimodular protein of around 600 kDa inhibits endothelial cell proliferation, migration and morphogenic organization into capillary tubes[39]. The role of thrombospondin-1 as an endogenous angiostatic molecule is supported by the phenotype of the null mice, which exhibit excessive vascularity in the dermis and pancreatic islets[40].

More recently, using the antiangiogenic motifs in thrombospondin-1, METH1/ADAMTS1, a secreted metalloprotease with three thrombospondin-1 motifs, has been cloned and shown to inhibit endothelial cell proliferation in vitro and block the neovascular response induced by growth factors in vivo. This is due to the fact that METH1/ADAMTS1 binds to VEGF and attenuates VEGFR-2 phosphorylation[41]. Also interferons, secreted glycoproteins initially characterized for their antiviral effects, are angiostatic, since they inhibit endothelial migration and downregulate the synthesis of basic FGF[42,43]. Because the ECM is an essential component of the angiogenic response, it is not surprising that tissue inhibitors of metalloproteases (TIMP) have been found to inhibit angiogenesis. Indeed, TIMPs suppress matrix metalloproteinase (MMP) activity critical for ECM turnover. However, TIMPs function also independent of MMP inhibition. TIMP-2 abrogates angiogenic factor-induced endothelial cell proliferation[44] and TIMP-3 blocks the binding of VEGF to VEGFR-2 thus inhibiting downstream signalling[45].

While some endogenous inhibitors are expressed under physiological conditions, they can also be generated in association with tumor growth. Among these, numerous potent antiangiogenic molecules are proteolytic fragments of larger naturally occurring proteins, such as collagen and plasminogen. Angiostatin was the first member of this group of inhibitors to be described[46]. It is a 38 kDa internal fragment of plasminogen that potently inhibits capillary endothelial cell growth in vitro and neovascularization in tumor-bearing mice. Another member of this family is endostatin, a 20 kDa carboxy-terminal fragment of type XVIII collagen, which dramatically inhibits endothelial cell migration, vascular morphogenesis, and perivascular cell recruitment in vivo[47]. Canstatin and tumstatin originate from the same precursor, collagen IV. Canstatin, a 24-kDa peptide derived from the C-terminal globular non-collagenous domain of the $\alpha_2$ chain, blocks angiogenesis because it promotes endothelial apoptosis by inhibiting phosphatidylinositol 3-kinase/Akt[48]. Tumstatin, a 28-kDa derived from $\alpha_3$ chain, functions as an endothelial cell-specific inhibitor of protein synthesis through its interaction with $\alpha_V\beta_3$ integrin[49].

The role of chemokines in angiogenesis is intriguing since both pro-angiogenic chemokines, such as CXCL8/IL8 interacting with the CXCR2 receptor, and anti-angiogenic chemokines, such as CXCL10/IP10 interacting with the CXCR3 receptor, have been described[50].

Many other cofactors are involved. Interestingly, some molecules contributing to angiogenesis, such as Notch, Wnt/ βcatenin, Hedgehog, are shared by developmental processes, thus raising the question about the evolution of the mechanisms of angiogenesis. Also the concentration of trace elements in the tissue may play a role. Magnesium is required for endothelial growth and migration[51]. Indeed, it has been shown that low magnesium impairs angiogenesis[52]. Copper is a potent inducer of angiogenesis since it directly acts on endothelial cells stimulating their proliferation and migration and also induces the expression of VEGF[53]. On the contrary, selenium promotes endothelial apoptosis[54].

It is therefore clear that endothelial cells are targeted by a myriad of different molecules that must cooperate in a temporally and spatially coordinated manner to make the process successful.

# 3. ANGIOGENIC FACTORS AND INTRACELLULAR SIGNALING TRANSDUCTION PATHWAYS

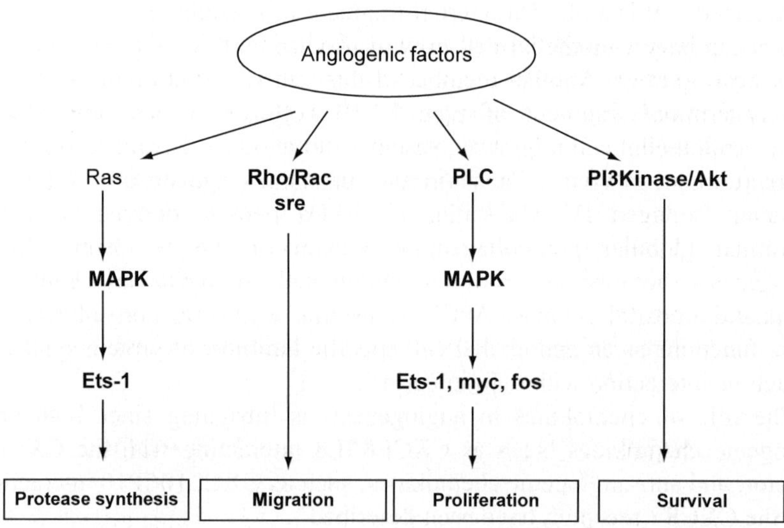


*Figure 1-2.* Principal intracellular signalling pathways involved in the regulation of some steps of angiogenesis.

The transition from a quiescent to an angiogenic phenotype in endothelial cells is governed by the activation of several signal transduction pathways as a response to specific stimuli (Figure 1-2). One of the earliest events in angiogenesis is the upregulation of endothelial proteolytic enzymes which degrade the basal lamina underneath the endothelium. The Ras-MAPK pathway is pivotal because it targets the zinc-finger transcription factors Ets-1 which plays a key role in activating the proteolytic system. After the matrix degradation by proteases, the endothelial cell migration occurs. A key role is played by small GTPase and cytosolic tyrosin kinases which must be activated to stimulate the formation of focal adhesions, a pre-requisite for endothelial migration. Indeed, VEGFR-2 activates the Focal Adhesion Kinase (FAK) that regulates the organization of the cytoskeleton and recruits other molecules, including the tyrosine kinase Src. In short, signaling molecules associate with focal contacts. The small GTPases Rho and Rac control cytoskeletal dynamics as well as polarization and migration. Following migration, endothelial cells begin to proliferate, an event mainly mediated by the extracellular signal-regulated kinase mitogen-activated protein kinases (ERK-MAPK) cascade, which is activated by different signal transduction pathways including phospholipase C gamma and the small GTPase Ras. VEGFR-2 and FGF receptors, with the contribution of integrins, activate these signal transducers. This leads to the translocation of the activated ERK-MAPK to the nucleus and the induction of several transcription factors involved in cell proliferation, among which are c-Fos, c-Myc, Ets-1.

After migrating and proliferating, endothelial cells should return to quiescence, survive, undergo a morphogenetic program and recruit perivascular cells. This is the convergence point for integrin-mediated interactions with the ECM and soluble factors such as VEGF and FGF. Although the signalling mechanisms controlling these late steps are less known, the p38 MAPK and JNK pathways seem involved. The PI3kinase/Akt pathway is mainly responsible for endothelial survival. From this rapid overview, it is clear that the regulation of endothelial behaviour in angiogenesis is the result of a very complex network of intracellular signalling systems that trigger, control and terminate the process.

## 4. ANGIOGENESIS AS A MULTISTEP PROCESS

Developing, injured or diseased tissues release angiogenic growth factors that diffuse into the nearby tissues, bind to specific receptors located on the endothelial cells of nearby preexisting blood vessels and activate endothelial cells. This is the beginning of a complex series of events which include: i) an initiation phase, characterized by endothelial cell shape changes, retraction of pericytes, destabilization of the vessel and increased permeability; ii) a progression phase, constituted by the production of proteolytic enzymes that degrade the ECM, endothelial cell migration and proliferation; iii) a differentiation phase, characterized by stabilization, remodeling and maturation of the vessels.

*Initiation.* In quiescent vessels, endothelial cells establish and maintain a permeability barrier through cell-cell adhesion molecules, which serves both a mechanical and a signaling function, contributing to growth inhibition and survival[55]. Tight junction molecules (claudin, occludin, jam-1), adherens junction molecules (VE-cadherin), gap junction molecules (connexin-37, -40, -43), PECAM-1/CD31, as well as a molecular crosstalk between endothelial and neighboring mural cells are crucial in regulating vessel stability and permeability. In the beginning, the endothelial cells must pick up the moorings from the mother vessel. This event is mainly attributed to VEGF. Hypoxia and nitric oxide (NO) induce the synthesis of VEGF[56]. In turn, the generated VEGF increases vascular permeability by inducing additional NO and prostacyclin[57] and through destabilizing cell-cell adhesion molecules. Indeed, VEGF stimulates the phosphorylation of β- and γ catenins as well as p120-catenin, which are intracellular messengers of VE-cadherin. Consequently, VE-cadherin fails to cluster at the cell-cell contacts and redistributes along the cell surface[58]. In the early steps of angiogenesis, Ang-2 levels increase contributing to the destabilization by blocking Tie-2 signalling and promoting endothelial cell-cell junction rearrangement and perivascular cell detachment. The increase of permeability causes extravasation of plasma proteins, allowing the formation of a provisional matrix of fibrin which will facilitate endothelial migration.

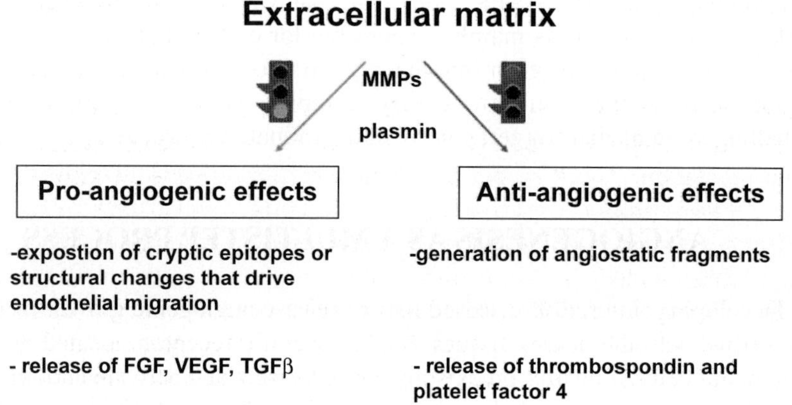


*Figure 1-3.* Remodelling of the extracellular matrix by proteases can have either pro- or anti-angiogenic effects.

*Progression.* In quiescent vessels, a basement membrane of collagen IV, laminin and other components encases endothelial cells and pericytes. Extracellular proteolysis by matrix metalloproteases, aminopeptidases or other neutral proteases is an absolute requirement for angiogenesis[59,60]. In addition, it emerges that transmembrane proteases such as ADAM-17 and MT1-MMP might also participate in the angiogenic process[61]. Activated endothelial cells locally degrade this matrix network by dissolving tiny holes in the basement membrane and in the interstitial stroma to allow endothelial cell migration in the perivascular space towards the chemotactic angiogenic stimuli. Temporal and spatial regulation of ECM remodelling events also contributes to the control of cell growth and migration. Remodelling of the ECM can have either pro- or anti-angiogenic effects (Figure 1-3) since cryptic segments or pre-existing domains within larger proteins of the matrix components can be exposed by conformational changes and/or generated by partial enzymatic hydrolysis. These exposed domains can positively or negatively regulate important functions of endothelial cells including adhesion, migration, proliferation, cell survival and cell-cell interactions. In addition, proteases can facilitate endothelial sprouting by liberating matrix-bound angiogenic factors, among which are VEGF and FGF. In parallel, new matrix synthesized by stromal cells is laid down. This new matrix, coupled with soluble growth factors, fosters the migration of endothelial cells by orchestrating cell-matrix and cytoskeletal rearrangements necessary for the cell motility. These signals are also relevant in driving endothelial proliferation. Under the action of VEGF and other angiogenic molecules, the endothelial cells multiply, loosely following each other into the perivascular space and forming a migration column which leads to a differentiation zone where endothelial cells change shape, adhere to each other and begin to organize to form a lumen.

*Differentiation, stabilization, remodeling and maturation of vessels.* Although endothelial cell differentiation is recognized as an important component of angiogenesis, little is known about the molecular mechanisms of its regulation. *In vitro*, the endothelial cell is capable of activating a unique genetic program in response to environmental signals that direct and sustain the formation of a differentiated phenotype. Cytokines and ECM components, including fibrin and collagen, induce changes from the traditional nonpolar cobblestone monolayer into a polar elongated fibroblast-like phenotype[62]. These polar endothelial cells ultimately organize into three-dimensional capillary-like structures[63]. Endothelial cell differentiation has a transcriptional basis since it is associated with an increase in the transcripts encoding fibronectin[64], the protein G-coupled receptor EDG-1[65], the receptor for sphingosine 1-phosphate, Jagged, the ligand for the Notch receptor[66], and EDF-1, a calmodulin-binding protein which also acts as a transcriptional coactivator[67]. Endothelial cell differentiation is also associated with a decrease in *sis*-mRNA[64]. Once activated this genetic program, endothelial multicellular aggregates align, organize, and result in the formation of stalks and finally form multicellular capillary-like structures. Cell-cell contacts must then be reconstructed to maintain a physical connection between endothelial cells, but also to promote their organization during lumen formation. Adhesion molecules such as PECAM and VE-cadherin are important during the early steps of vessel assembly[68]. PECAM is a regulator of vessel permeability and contributes to maintain the integrity of the endothelial barrier. A new angiogenic vessel expresses and distributes PECAM at cell-cell contacts to increase physical connections and reduce permeability. Similarly, VE-cadherin stimulates the stabilization of endothelial cell connections as well as lumen formation. Knock-out (KO) mice for VE-cadherin die 9.5 days after implantation showing several vascular defects including a remarkable reduction of the lumen in many vessels[55]. Also $\alpha_v\beta_3$ and $\alpha_v\beta_5$ integrins contribute to lumen formation, while trombospondin-1 inhibits it. Once the vascular lumen has been established, the new vessels move toward the remodelling phase, in which they complete the network organization, functional specification, and structural stabilization. In the so called branching phase, collateral vessels are created by sprouting, bridging and intussusception[68]. Some isoforms of VEGF are involved in branching. Indeed, the excision of exons 6 and 7 of the VEGF gene to generate mice expressing only $VEGF_{120}$ leads to ischemic heart disease with severe defects of tissue perfusion[69]. Ang1 and its receptor Tie2 influence the microvascular intussusceptive growth[70]; accordingly, KO mice exhibit dilated vessels with poor ramifications. Local factors should be taken into account since the newly formed vascular network must meet the needs of the specific tissues in which they are growing. A tissue specific function is described for renin in the branching of renal arteries[71] and for acidic FGF in the ramification of myocardial arteries[72].

The structural stabilization of new angiogenic vessels include the recruitment of perivascular cells which are important not only as mechanic coating for the endothelial layer but also as regulatory components of the vessel growth. A complex interplay exists between the endothelium and the mural cells. Endothelial cells secrete soluble factors, among them is PDGF-BB, which are involved in differentiation, recruitment, and adhesion to the vessel wall of smooth muscle cells (SMC) and pericytes[73]. Once perivascular cells reach the new vessel, they participate in the stabilization of the vessel architecture by releasing soluble factors. In particular, Ang-1 tightens the interactions between endothelial and perivascular cells and inhibits the VEGF-dependent dissociation of β-catenin from VE-cadherin, thus stabilizing endothelial cell-cell contacts[74]. Indeed, mutations of the Tie-2 receptor cause the loss of SMC and vascular malformation in humans[74]. Angiopoietins and their receptors don't work alone. Members of the TGF-β family such as TGFβ1, TGFβ-R2, endoglin, Smad5, play a role in SMC differentiation, ECM deposition and inhibition of endothelial cell migration and proliferation[75]. Many other molecules regulating mesenchimal-endothelial cross-talk must be at least cited, such as integrins, endothelin-1, Ephrin-Eph, dHAND and N-cadherin, among others[9,69]. The "baby" angiogenic vessel, once assembled, receives blood flow from the mother vessel. The lumen diameter must be appropriately regulated in order to sustain the blood pressure and this happens as the result of a balance between the action of lumen-increasing and lumen-decreasing factors. $VEGF_{165}$ and $VEGF_{121}$ together with Ang1 increase lumen diameter, whereas $VEGF_{189}$ decreases lumen diameter[76]. Recently, new genes have been shown to participate to lumen regulation among which tubedown-1 (tbd-1), a novel acetyltransferase which may participate with TGF-β2 in diminution of blood vessel development[77]. Now that active blood flow is established, the late step of vascular remodelling occurs. Indeed, blood pressure and shear stress induce expression of many endothelial and SMC genes such as c-Fos, Egr-1, ACE, eNOS, PDGF-A and B, TGF-β and integrins[78] which contribute to the maturation of the vessels. At the end, the endothelial cells go back into quiescence and a long term survival program is activated by Ang-1/Tie-2 and VEGF/VE-cadherin. Briefly, Ang-1 enhances survivin expression via akt[79]. Concomitantly, VEGF activates its receptor to cluster with VE-cadherin, PI3-kinase and β-catenin. The signal trasduction pathway initiated by the multiprotein complex leads to the upregulation via akt of bcl-2, an anti-apoptotic oncogene[55]. Also mechanical factors influence survival since shear stress reduces endothelial cells turnover and inhibits TNFα dependent apoptosis[80].

Once the vessel is mature, the endothelial cells become remarkably resistant to exogenous factors and are quiescent. In addition, local physiological requirements must be met and endothelial cells may acquire highly specialized characteristics. For instance, tight junctions are particularly impermeable in the brain, while endothelial cells in endocrine glands are discontinuous and fenestrated. The factors that regulate the acquisition of these specific properties are largely unknown.

## 5. ANGIOGENESIS IN HEALTH AND DISEASE

In a physiological setting, angiogenesis is confined to the metabolic demands of growing or healing tissues. The high coordination of the process is demonstrated in cutaneous wound healing. Rapidly after a wound has been afflicted, Ang1 is downregulated while VEGF and Ang2 are overexpressed with similar kinetics, thus allowing the destabilization of the existing vessels followed by the formation of new capillaries. Later on, VEGF and Ang2 decrease to the baseline levels so that the new vessels rapidly mature and become stable. The entire process is finely tuned in a specific temporal sequence that results in a tightly regulated balance between pro- and anti-angiogenic signals[81]. Following an analogous program, neovascularization occurs after a bone fracture and permits not only the resorption of necrotic tissue but also the introduction of osteogenic stem cells.

Similar events and molecular interplay are described in the cyclic angiogenesis that occurs uniquely within the female reproductive tract and is critical for normal reproduction[82]. In particular, the corpus luteus grows and vascularizes extremely rapidly. Again, VEGF is major player. Luteal expression of VEGF occurs primarily in perivascular cells and is regulated by oxygen levels. Moreover, nitric oxide, which is a potent vasodilator and can stimulate VEGF production, is synthesized by the endothelial cells of luteal arterioles and capillaries. Therefore a paracrine loop exists between the vascular endothelial cells, which produce NO, and the peri-endothelial cells (vascular smooth muscle and pericytes), which produce VEGF, to ensure a coordinated regulation of luteal vasodilation and angiogenesis[83].

Certain pathological conditions usurp the aforementioned regulatory mechanisms, being tumor angiogenesis the most characterized example. Physiological and pathological angiogenesis, while sharing most molecular mechanisms, are not totally alike. Evidence is emerging that, apart from the aforementioned pro- and anti-angiogenic factors, different molecules - such as cycloxygenase-2, Placental Growth Factor (PlGF), proteases and thrombospondin-2 are implicated in pathological neovascularization[84-86]. Another difference between physiological and pathological angiogenesis is that the latter is often determined by inflammation. As mentioned above, monocyte/macrophages or other leukocytes infiltrate the diseased tissues and secrete angiogenic factors that, in turn, attract endothelial and other cells contributing to the formation of new vessels.

## 5.1 Tumor angiogenesis

Tumors, which have been intriguingly described as "wounds that never heal"[87], have lost the appropriate balance between positive and negative regulators of angiogenesis. Neovascularization sustains tumor growth because it nourishes tumor cells and because endothelial cells stimulate malignant cell growth in a paracrine manner. Moreover, angiogenesis coincides with increased tumor cell entry into the circulation facilitating metastatic spread. Several lines of evidence indicate that the angiogenic process precedes the formation of the malignant tumor, suggesting that angiogenesis may represent the rate-limiting step not only for tumor expansion but also for the onset of malignancy[88]. Interestingly, tumor vessels develop unique characteristics that render them distinct from the normal vasculature and generate a chaotic and variable blood flow so that some regions of the tumor may become hypoxic. Tumor vessels are irregularly shaped, dilated, tortuous, with uneven diameter and excessive branching sometimes leading to dead ends. They are often leaky, with numerous openings, widened interendothelial junctions and discontinuous basal membrane thus demonstrating increased permeability to circulating large molecules up to sizes of 400 nm in diameter[89]. This is partly due to an imbalance between VEGF and angiopoietins. The endothelial cells are often abnormal in shape, eventually growing on the top of each other. In addition, tumor vessels lack functional perivascular cells and their wall is not always lined by endothelial cells, but also by cancer cells alone or by a mosaic of cancer and endothelial cells. Until recently, however, endothelial cells within the tumor were considered genetically stable, and were, therefore, proposed as a good target for therapy. Recently, a varying number of the tumor-associated vascular endothelial cells have been shown to possess identical chromosomal aberrations to those of the tumor itself, thus indicating that targeting endothelial cells may be more complicated that originally thought[90].

The structural and pathological abnormalities in tumor vessels reflect the pathological nature of their induction and differ among tumor types. Briefly, during sprouting angiogenesis, vessels become leaky in response to VEGF released by tumor and inflammatory cells. Ang2 mediates the dissolution of endothelial junctions and proteases digest the basement membrane and the interstitial matrix. Many different molecules stimulate endothelial proliferation, migration and assembly, among which, VEGF, FGFs, Ang1, PlGF, and certain integrins. In parallel, most of the inhibitors are downregulated. This uncoordinated and chaotic situation leads to the formation of immature and dysfunctional vessels.

Angiogenesis is not the only mechanism utilized by tumors to nourish themselves, although it is always present. Indeed, some tumor vessels can form also by longitudinal division of existing vessels with periendothelial cells, a process known as intussusception. Alternatively, tumor angiogenesis can occur by incorporation of bone marrow derived endothelial precursors or by coopting existing vessels[91]. The typical example is astrocytoma which first acquires blood supply by coopting existing normal brain blood vessels, which means that astrocytoma grows along the vessels. However, when the tumor becomes hypoxic partly because of an increased neoplastic cell proliferation, angiogenesis is initiated to supply the tumor with the necessary metabolites[92].

## 5.2 Excessive angiogenesis and non neoplastic diseases

Exuberant angiogenesis occurs at anytime when the diseased cells produce abnormal amounts of angiogenic growth factors, overwhelming the effects of natural angiostatic molecules. It is a pathogenic step in many common diseases: rheumatoid arthritis, retinopathies, atherosclerosis, neurodegeneration, obesity, among others. Rheumatoid arthritis (RA) is one of the first diseases in which neovascularization has been shown to have a pathogenic role. The expansion of the synovial lining of joints in RA and the subsequent invasion by the pannus of the underlying cartilage and bone necessitates an increase in the vascular supply to the synovium in order to cope with the increased requirement for oxygen and nutrients. Although many pro-angiogenic factors are expressed in the synovium in RA, VEGF has a central involvement in the angiogenic process. It is found in the synovial fluid and serum of patients with RA and its expression positively correlates with disease severity. The relevant role of VEGF is underscored by the fact that its inhibition by synthetic compounds, such as TNP-470, or by naturally occurring factors, such as the soluble VEGF receptor, produces therapeutic effects[93].

Exuberant angiogenesis is a leading cause of visual impairment and blindness. Indeed, common ocular diseases such as diabetic retinopathy, retinopathy of prematurity, neovascular glaucoma, and age-related macular degeneration are all characterized by the breakdown of blood retinal barrier and neovascularization initiated and perpetrated mainly by dysregulated VEGF expression[94]. In particular, since retinal capillary coverage with pericytes is crucial for endothelial survival, it is noteworthy that pericyte loss is an early pathologic feature of diabetic retinopathy. It is now clear that chronic hyperglycemia upregulates Ang-2 in the diabetic retina and that Ang-2 overexpression is causally involved in pericyte loss[95]. Endothelial cells subsequently disappear, leaving behind acellular capillaries, which are no longer perfused. This stimulates proliferative retinopathy through the induction of VEGF.

Also some dermatologic diseases are angiogenesis-dependent. Kaposi's sarcoma, for instance, is a highly vascularized skin lesion in which dysregulated angiogenesis results from an interplay among inflammatory cytokines, angiogenic factors and viral agents[96]. The picture in psoriasis, a chronic inflammatory dermatosis, is peculiar, since non-sprouting angiogenesis leads to increased elongation and widening of dermal vessels, thus causing hypervascularity. Specifically, in psoriasis, hypervascular skin lesions overexpress the angiogenic chemokyne IL-8 and underexpress some inhibitors of angiogenesis[97].

It is also worth noting that atherosclerosis, the first cause of mortality in western world, involves plaque angiogenesis. Normally, the arterial intima is devoid of blood vessels, with oxygen and nutrients being supplied from the lumen and by the adventitial vasa vasorum. However, during plaque development, the intima, and to a lesser extent the media, become thickened and this is accompanied by the appearance of numerous blood vessels. It is again VEGF that promotes plaque development[98].

The role of angiogenesis in obesity is rather intriguing. Adipose tissue consists of adipocytes and vascular endothelial cells, which provide oxygen and nutrients to the growing mass. Interestingly, adipogenesis and angiogenesis are spatially and temporally coupled during embryonic development. VEGF, FGF and leptin have been identified as being mediators of angiogenesis in this context. Since angiogenesis appears to play a critical role in adipose tissue growth and reduction also in adulthood, it is not surprising that adipose tissue mass is sensitive to angiogenesis inhibitors, which can potentially serve to regulate overweight and obesity[99].

A last issue to consider is the role of angiogenesis in neurodegeneration. Neovascularization in the brain occurs after different insults that induce neuronal loss. Hypoxia, hypoglycemia, or inflammation are the triggers for the angiogenic response. When neurons are diseased, reactive glial cells secrete local factors that induce angiogenesis, probably as part of a spontaneous neuroprotective mechanism related to the increased metabolic demand. In Parkinson's disease, the increased vascularization seems to be induced, or at least perpetuated, by dopaminergic cell loss, which signals neighboring cells to release factors involved in neovascularization. The new vessels facilitate the entrance into the brain parenchyma of neurotoxins and harmful cytokine-releasing blood cells, both of which contribute to neuronal cell death[100]. Also Alzheimer's disease is linked with angiogenesis, which is induced in response to impaired cerebral perfusion and inflammatory mediators. Invading macrophages release VEGF and FGF, while thrombospondin expression near focal disease lesions is reduced, thus leading to a proangiogenic state. In this disease, the brain endothelium contributes to neurodegeneration because it secretes the precursor substrate for the β-amyloid plaque and a neurotoxic peptide that kills cortical neurons[101].

While exuberant angiogenesis is a common denominator in all these conditions, impaired angiogenesis is also cause of disease (Figure 1-4).

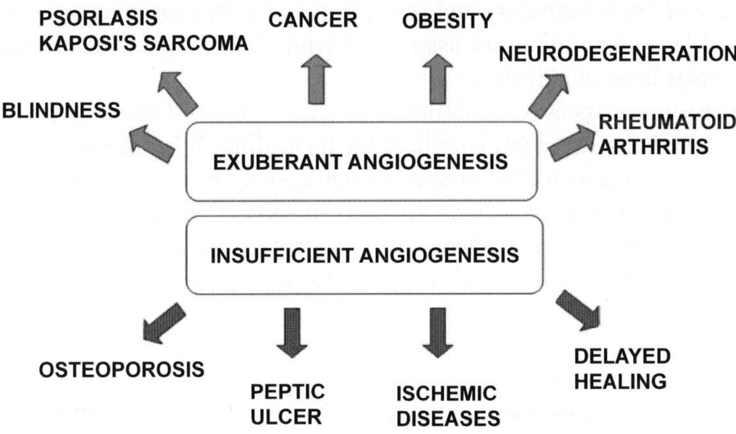


*Figure 1-4.* Pathologic angiogenesis and disease.

## 5.3 Insufficient angiogenesis and disease

The awareness that angiogenesis is pertinent in the context of cardiovascular disease has arisen from the fact that occlusion or narrowing of arteries results in hypoxia, in response to which the ischemic myocardium develops collateral vessels. However, this compensatory angiogenesis is often insufficient. In animal models, coronary artery occlusion stimulates VEGF expression and the formation of collateral blood vessels that circumvent the occlusion, indicating VEGF-mediated neovascularization as a critical adaptive response to chronic myocardial ischemia. In humans, the angiogenic response is usually insufficient and necrosis may occur. A defect in VEGF production sometimes reflects an age-dependent impairment in the induction of HIF-1 DNA-binding activity and VEGF gene transcription in response to hypoxia. However, patients of the same age and degree of coronary occlusion vary markedly in the extent of coronary collateralization. One potential explanation is the existence of individual variation with regard to the expression of the HIF-1/VEGF pathway in response to hypoxia or ischemia. The concept of therapeutic angiogenesis exploits and supplements the physiological response to hypoxia or ischemia. A number of approaches have been studied, with varying degrees of success[102].

On a similar basis, inadequate or inappropriate vascularity in the bone is associated with decreased bone formation and mass. It is noteworthy that the cytokines and growth factors regulating intraosseous angiogenesis also regulate bone remodelling, and close links exist between the blood supply to the bone and bone formation and resorption: most diseases characterized by decreased bone resorption are associated with altered bone vascularization. Osteoporosis is among them.

Peptic ulcers appear to be deficient in microvessels. Once a peptic ulcer has developed, angiogenesis is critical for its healing. It is noteworthy that in human gastric ulcers an impairment in angiogenesis has been reported and it associates with a marked downregulation of basic FGF. In addition, it has recently been shown that Helicobacter pylori, the etiologic agent of gastric ulcer, downregulates VEGF and Ang receptor in endothelial cells[103].

*Table 1-2.* Angiostatic drugs.

| Drug | Mechanism |
|---|---|
| Drugs that block matrix breakedown | |
| Marimastat (synthetic) | inhibition of MMP |
| Neovastat (natural) | inhibition of MMP |
| Drugs that inhibit endothlial cells | |
| TNP470 | inhibition of endothelial growth |
| Thalidomide | inhibition of endothelial response to VEGF and of ciloxygenase 2 expression |
| Squalamine | inhibition of sodium-hydrogen exchanger |
| Combrestatin | inhibition of apoptosis of proliferating endothelium |
| Endostatin | inhibition of endothial migration and vascular morphogenesis |
| Drugs that target angiogenic factor | |
| Monoclonal antibodies aginst VEGF | block of VEGF |
| SU5416 | block of VEGF receptor signalling |
| Interferons | inhibition of FGF and VEGF production |
| Drugs with non specific mechanisms of action | |
| Interleukin 12 | upregulation of interferon gamma e IP-10 |
| CAI | inhibition of calcium influx |

# 6. HOW TO MODULATE ANGIOGENESIS

A great deal of effort is now being devoted to the development of new drugs that hopefully will control pathological angiogenesis. Indeed, anti-angiogenic agents are currently used in the clinics and can be categorized into several classes based on the biological activity of the compounds used (Table 1-2). This area of clinical research is in constant evolution and the U.S. National Cancer Institute maintains an up-to-date website (http://www.cancer.gov). However, there is growing evidence that certain practical nutritional measures have the potential to slow angiogenesis and that bioactive compounds from natural sources may be used as regulatory agents. It is therefore reasonable to anticipate that, by combining several measures that work in distinct but complementary ways, it may be possible to control exaggerated angiogenesis.

A long list of dietary factors strongly inhibits blood vessel growth, among them resveratrol in red wine, as well as genistein in soya, catechins in green tea and brassinin in Chinese cabbage. Resveratrol, a polyphenolic antiangiogenic compound, inhibits intracellular signalling induced by growth factors. Its antiangiogenic mechanisms involve direct inhibition of capillary endothelial cell growth via suppression of the phosphorylation of the mitogen-activated kinase[104]. These effects are observed at concentrations that are likely to be achieved in blood after moderate wine consumption. Consequently, resveratrol inhibits angiogenesis-dependent physiological and pathological processes including wound healing and tumor growth. With similar molecular mechanisms, green tea polyphenols can suppress endothelial responsiveness to both VEGF and FGF.

Apart from preventing atherogenesis, supplementation with fatty omega-3-rich fish oil seems to be useful in preventing angiogenesis, since omega-3 fatty acids inhibit endothelial expression of VEGFR-2 and can also suppress tumor production of pro-angiogenic eicosanoids.

Flavonoids have been shown to inhibit angiogenesis and proliferation of endothelial cells in vitro. Genistein, a naturally occurring isoflavonoid, blocks the activity of tyrosine kinases, among which the receptors of angiogenic factors, and MMPs[105]. In addition, it downregulates VEGF expression through inhibiting hypoxic activation of HIF-1[106].

Some trace elements are beneficial in controlling angiogenesis. Copper ions stimulate endothelial proliferation and migration and also activate several angiogenic factors, among which VEGF and FGF. In light of evidence that angiogenesis has a high requirement for copper, copper depletion may have exceptional potential as an angiostatic measure, and is most efficiently achieved with the copper-chelating drug tetra-thiomolybdate[107].

Selenium has been shown to induce endothelial apoptosis principally through the p38 MAPK pathway[108]. It also inhibits endothelial expression of matrix metalloproteinase-2 and the expression of VEGF in cancer cells. The concentrations generating half-maximal inhibition of the aforementioned events are within the plasma range of selenium in US adults. All together, these effects may explain the chemopreventive effects of this trace element. Supplementation with selenium may therefore be physiologically pertinent to control the angiogenic switch in early lesions by retarding and, possibly, blocking their growth and progression. It may be helpful also in all those situations characterized by excessive angiogenesis.

Should we need any further indication of the health hazard of smoking, it is now clear that cigarette smoke impairs angiogenesis. Indeed, smoking significantly impairs blood flow recuperation (laser-Doppler imaging) in a model of hind-limb ischemia, an observation that correlates with an important reduction of capillary density in ischemic muscles. At the cellular and molecular level, cigarette smoke exposure impairs HIF-1$\alpha$ stabilization and accumulation under hypoxic conditions, leading to down-regulation of hypoxia-induced VEGF expression, which in turn results in reduced angiogenesis in response to hypoxia[109]. The exact components of cigarette smoke responsible for the impairment of angiogenesis are unknown. Cigarette smoke is composed of nearly 4000 different chemicals, many of which (carbon monoxide, cadmium, hydrocarbons, acetaldehyde, etc.) are toxic to endothelial cells and detrimental to health. Carbon monoxide has been shown to destabilize HIF-1$\alpha$ and suppress the activation of target genes. Nicotine, on the other hand, was shown to promote angiogenesis in different models. Nonetheless, the cumulative effect of the different components of cigarette smoke has a negative effect on angiogenesis with a major impact in patients with ischemia and peptic ulcers, conditions that require a strict non-smoking regimen.

## 7. CONCLUDING REMARKS

The insights into the molecular basis of angiogenesis have resulted in treatment paradigms to modulate angiogenesis. However, a long list of puzzling questions remains unanswered. Although in the global database PubMed/Medline more than 21000 articles can be found using the keyword "angiogenesis", 14000 being published in the last five years, more work is needed in the future, but, in general, the outlook has become a promising one.

# REFERENCES

1. Hunter J. Treatise on the blood, inflammation and gunshot wounds. Philadelphia, Thomas Bradford, 1794.
2. Egginton S. Temperature and angiogenesis: the possible role of mechanical factors in capillary growth. Comp Biochem Physiol. A Mol Integr Physiol. 2002, 132 (4): 773-787.
3. Folkman J. Successful treatment of an angiogenic disease. N Engl J Med. 1989, 320 (18): 1211-1212.
4. Carmeliet P. Angiogenesis in health and disease. Nat Med. 2003, 9 (6): 653-660.
5. Cines DB, Pollak ES, Buck CA, Loscalzo J, Zimmerman GA, McEver RP, Pober JA, Wick TM, Konkle BA, Schwartz BS, Barnathan ES, McCrae KR, Hug BA, Schmidt AM, Stern DM. Endothelial cells in physiology and in the pathophysiology of vascular disorders. Blood 1998, 91: 3527-3561.
6. Hanahan D, Folkman J. Patterns and emerging mechanisms of the angiogenic switch during tumorigenesis. Cell 1996, 86 (3): 353-364.
7. Hirschi KK, D'Amore PA. Pericytes in the microvasculature. Cardiovasc Res. 1996, 32 (4): 687-698.
8. Reinmuth N, Liu W, Jung YD, Ahmad SA, Shaheen RM, Fan F, Bucana CD, McMahon G, Gallick GE, Ellis LM. Induction of VEGF in perivascular cells defines a potential paracrine mechanism for endothelial cell survival. FASEB J. 2001, 15 (7): 1239-1241.
9. Jain RK. Molecular regulation of vessel maturation. Nat Med. 2003, 9 (6): 685-693.
10. Silvakumar B, Harry LE, Paleolog EM. Modulating angiogenesis: more vs less. JAMA 2004, 292: 972-977.
11. Krogh A. The number and distribution of capillaries in muscles with calculations of the oxygen pressure head necessary for supplying the tissue. J Physiol. 1919, 52: 409-415.
12. Pugh CW, Ratcliffe PJ. Regulation of angiogenesis by hypoxia: role of the HIF system. Nat Med. 2003, 9 (6): 677-684.
13. Ivan M, Kondo K, Yang H, Kim W, Valiando J, Ohh M, Salic A, Asara JM, Lane WS, Kaelin WG Jr. HIFalpha targeted for VHL-mediated destruction by proline hydroxylation: implications for O2 sensing. Science 2001, 292 (5516): 464-468.
14. Knighton DR, Hunt TK, Scheuenstuhl H, Halliday BJ, Werb Z, Banda MJ. Oxygen tension regulates the expression of angiogenesis factor by macrophages. Science 1983, 221 (4617): 1283-1285.
15. Kourembanas S, Hannan RL, Faller DV. Oxygen tension regulates the expression of the platelet-derived growth factor-B chain gene in human endothelial cells. J Clin Invest. 1990, 86 (2): 670-674.
16. Pichiule P, Chavez JC, LaManna JC. Hypoxic regulation of angiopoietin-2 expression in endothelial cells. J Biol Chem. 2004, 279 (13): 12171-12180.
17. Tang N, Wang L, Esko J, Giordano FJ, Huang Y, Gerber HP, Ferrara N, Johnson RS. Loss of HIF-1alpha in endothelial cells disrupts a hypoxia-driven VEGF autocrine loop necessary for tumorigenesis. Cancer Cell. 2004, 6 (5): 485-495.
18. Risau W. Mechanisms of angiogenesis. Nature. 1997, 386 (6626): 671-674.
19. Reynolds LP, Redmer DA. Expression of the angiogenic factors, basic fibroblast growth factor and vascular endothelial growth factor, in the ovary. J Anim Sci. 1998, 76 (6): 1671-1681.
20. Vacca A, Ribatti D, Iurlaro M, Albini A, Minischetti M, Bussolino F, Pellegrino A, Ria R, Rusnati M, Presta M, Vincenti V, Persico MG, Dammacco F. Human lymphoblastoid cells produce extracellular matrix-degrading enzymes and induce endothelial cell

proliferation, migration, morphogenesis, and angiogenesis. Int J Clin Lab Res. 1998, 28 (1): 55-68.

21. Norrby K. Mast cells and angiogenesis. APMIS. 2002, 110 (5): 355-371.
22. Melder RJ, Koenig GC, Witwer BP, Safabakhsh N, Munn LL, Jain RK. During angiogenesis, vascular endothelial growth factor and basic fibroblast growth factor regulate natural killer cell adhesion to tumor endothelium. Nat Med. 1996, 2 (9): 992-997.
23. Bernardini G, Ribatti D, Spinetti G, Morbidelli L, Ziche M, Santoni A, Capogrossi MC, Napolitano M. Analysis of the role of chemokines in angiogenesis. J Immunol Methods. 2003, 273 (1-2): 83-101.
24. Dermond O, Ruegg C. Inhibition of tumor angiogenesis by non-steroidal anti-inflammatory drugs: emerging mechanisms and therapeutic perspectives. Drug Resist Updat. 2001, 4 (5): 314-321.
25. Maxwell PH, Pugh CW, Ratcliffe PJ. Activation of the HIF pathway in cancer. Curr Opin Genet Dev. 2001, 11: 293-299.
26. Bikfalvi A, Bicknell R. Recent advances in angiogenesis, anti-angiogenesis and vascular targeting. Trends Pharmacol Sci. 2002, 23 (12): 576-582.
27. Saharinen P, Petrova TV. Molecular regulation of lymphangiogenesis. Ann N Y Acad Sci. 2004, 1014: 76-87.
28. Klagsbrun M, Takashima S, Mamluk R. The role of neuropilin in vascular and tumor biology. Adv Exp Med Biol. 2002, 515: 33-48.
29. Yancopoulos GD, Davis S, Gale NW, Rudge JS, Wiegand SJ, Holash J. Vascular-specific growth factors and blood vessel formation. Nature. 2000, 407: 242-248.
30. Carmeliet P, Ferreira V, Breier G, Pollefeyt S, Kieckens L, Gertsenstein M, Fahrig M, Vandenhoeck A, Harpal K, Eberhardt C, Declercq C, Pawling J, Moons L, Collen D, Risau W, Nagy A. Abnormal blood vessel development and lethality in embryos lacking a single VEGF allele. Nature. 1996, 380 (6573): 435-439.
31. Ferrara N, Chen H, Davis-Smyth T, Gerber HP, Nguyen TN, Peers D, Chisholm V, Hillan KJ, Schwall RH. Vascular endothelial growth factor is essential for corpus luteum angiogenesis. Nat Med. 1998, 4 (3): 336-340.
32. Larcher F, Murillas R, Bolontrade M, Conti C, Jorcano JL. VEGF overexpression in skin of transgenic mice induces angiogenesis, vascular hyperpermeability and accelerated turnover development. Oncogene. 1998, 17: 303-311.
33. Thomas KS. Fibroblast Growth Factors. FASEB J. 1987, 1: 434-440.
34. Miller DL, Ortega S, Bashayan O, Basch R, Basilico C.Munoz-Chapuli R, Quesada Compensation by fibroblast growth factor 1 (FGF1) does not account for the mild phenotypic defects observed in FGF2 null mice. Mol Cell Biol. 2000, 20 (6): 2260-2268.
35. Munoz-Chapuli R, Quesada AR, Angel Medina M. Angiogenesis and signal transduction in endothelial cells. Cell Mol Life Sci. 2004, 61 (17): 2224-2243.
36. Reynolds LE, Wyder L, Lively JC, Taverna D, Robinson SD, Huang X, Sheppard D, Hynes RO, Hodivala-Dilke KM. Enhanced pathological angiogenesis in mice lacking beta3 integrin or beta3 and beta5 integrins. Nat Med. 2002, 8 (1): 27-34.
37. Papetti M, Herman IN. Mechanisms of normal and tumor-derived angiogenesis. Am J Physiol Cell Physiol. 2002, 282: C947-C970.
38. Surawska H, Ma PC, Salgia R. The role of ephrins and Eph receptors in cancer. Cytokine Growth Factor Rev. 2004, 15 (6): 419-433.
39. Vailhe B, Feige JJ. Thrombospondins as anti-angiogenic therapeutic agents. Curr Pharm Des. 2003, 9 (7): 583-588.

PDGF-B leads to pericyte loss and glomerular, cardiac and placental abnormalities. Development. 2004, 131 (8): 1847-1857.
74. Thurston G. Role of Angiopoietins and Tie receptor tyrosine kinases in angiogenesis and lymphangiogenesis. Cell Tissue Res. 2003, 314 (1): 61-68.
75. Lebrin F, Deckers M, Bertolino P, Ten Dijke P. TGF-beta receptor function in the endothelium. Cardiovasc Res. 2005, 65 (3): 599-608.
76. Suri C, McClain J, Thurston G, McDonald DM, Zhou H, Oldmixon EH, Sato TN, Yancopoulos GD. Increased vascularization in mice overexpressing angiopoietin-1. Science. 1998, 282 (5388): 468-471.
77. Gendron RL, Adams LC, Paradis H. Tubedown-1, a novel acetyltransferase associated with blood vessel development. Dev Dyn. 2000, 218 (2): 300-315.
78. Brown MD, Hudlicka O. Modulation of physiological angiogenesis in skeletal muscle by mechanical forces: involvement of VEGF and metalloproteinases. Angiogenesis. 2003, 6 (1):1-14.
79. Papapetropoulos A, Fulton D, Mahboubi K, Kalb RG, O'Connor DS, Li F, Altieri DC, Sessa WC. Angiopoietin-1 inhibits endothelial cell apoptosis via the Akt/survivin pathway. J Biol Chem. 2000, 275 (13): 9102-9105.
80. Risau W. Differentiation of the endothelium. FASEB J. 1995, 9 (10): 926-933.
81. Bloch W, Huggel K, Sasaki T, Grose R, Bugnon P, Addicks K, Timpl R, Werner S. The angiogenesis inhibitor endostatin impairs blood vessel maturation during wound healing. FASEB J. 2000, 14 (15): 2373-2376.
82. Jaffe RB. Importance of angiogenesis in reproductive physiology. Semin Perinatol. 2000, 24 (1): 79-81.
83. Reynolds LP, Grazul-Bilska AT, Redmer DA. Angiogenesis in the corpus luteum. Endocrine. 2000, 12 (1): 1-9.
84. Heymans S, Luttun A, Nuyens D, Theilmeier G, Creemers E, Moons L, Dyspersin GD, Cleutjens JP, Shipley M, Angellilo A, Levi M, Nube O, Baker A, Keshet E, Lupu F, Herbert JM, Smits JF, Shapiro SD, Baes M, Borgers M, Collen D, Daemen MJ, Carmeliet P. Inhibition of plasminogen activators or matrix metalloproteinases prevents cardiac rupture but impairs therapeutic angiogenesis and causes cardiac failure. Nat Med. 1999, 5 (10): 1135-1142.
85. Varner JA, Brooks PC, Cheresh DA. The integrin alpha V beta 3: angiogenesis and apoptosis. Cell Adhes Commun. 1995, 3 (4):367-374.
86. Murohara T, Asahara T, Silver M, Bauters C, Masuda H, Kalka C, Kearney M, Chen D, Symes JF, Fishman MC, Huang PL, Isner JM. Nitric oxide synthase modulates angiogenesis in response to tissue ischemia. J Clin Invest. 1998, 101 (11): 2567-2578.
87. Dvorak HF. Tumors: wounds that do not heal. Similarities between tumor stroma generation and wound healing. N Engl J Med. 1986, 315 (26): 1650-1659.
88. Bergers G, Benjamin LS Tumorigenesis and the angiogenic switch. Nat Rev Cancer. 2003, 3: 401-410.
89. Brown MJ, Giacca AJ The unique physiology of solid tumors: opportunities (and problems) for cancer therapy. Cancer Res. 1998, 58: 1408-1416.
90. Streubel B, Chott A, Huber D, Exner M, Jager U, Wagner O, Schwarzinger I. Lymphoma-specific genetic aberrations in microvascular endothelial cells in B-cell lymphomas. N Engl J Med. 2004, 351 (3): 250-259.
91. Carmeliet P, Jain RK. Angiogenesis in cancer and other diseases. Nature. 2000, 407 (6801): 249-257.

92. Holash J, Maisonpierre PC, Compton D, Boland P, Alexander CR, Zagzag D, Yancopoulos GD, Wiegand SJ. Vessel cooption, regression, and growth in tumors mediated by angiopoietins and VEGF. Science. 1999, 284 (5422): 1994-1998.
93. Paleolog EM. Angiogenesis in rheumatoid arthritis. Arthritis Res. 2002, 4 Suppl 3: S81-90.
94. Wilkinson-Berka JL. Vasoactive factors and diabetic retinopathy: vascular endothelial growth factor, cycoloxygenase-2 and nitric oxide. Curr Pharm Des. 2004, 10 (27): 3331-3348.
95. Hammes HP, Lin J, Wagner P, Feng Y, Vom Hagen F, Krzizok T, Renner O, Breier G, Brownlee M, Deutsch U. Angiopoietin-2 causes pericyte dropout in the normal retina: evidence for involvement in diabetic retinopathy. Diabetes. 2004, 53 (4): 1104-1110.
96. Ensoli B, Sgadari C, Barillari G, Sirianni MC, Sturzl M, Monini P Biology of Kaposi's sarcoma. Eur J Cancer. 2001, 37 (10): 1251-1269.
97. Nickoloff BJ, Mitra RS, Varani J, Dixit VM, Polverini PJ. Aberrant production of interleukin-8 and thrombospondin-1 by psoriatic keratinocytes mediates angiogenesis. Am J Pathol. 1994, 144 (4): 820-828.
98. Lip GY, Blann AD. Thrombogenesis, atherogenesis and angiogenesis in vascular disease: a new vascular triad. Ann Med. 2004, 36 (2): 119-125.
99. Rupnick MA, Panigrahy D, Zhang CY, Dallabrida SM, Lowell BB, Langer R, Folkman MJ. Adipose tissue mass can be regulated through the vasculature. Proc Natl Acad Sci U S A. 2002, 99 (16): 10730-10735.
100. Barcia C, Emborg ME, Hirsch EC, Herrero MT. Blood vessels and parkinsonism. Front Biosci. 2004, 9: 277-282.
101. Vagnucci AH Jr, Li WW. Alzheimer's disease and angiogenesis. Lancet. 2003, 361 (9357): 605-608.
102. Fam NP, Verma S, Kutryk M, Stewart DJ. Clinician guide to angiogenesis. Circulation. 2003, 108 (21): 2613-2618.
103. Kim JS, Kim JM, Jung HC, Song IS. Helicobacter pylori down-regulates the receptors of vascular endothelial growth factor and angiopoietin in vascular endothelial cells: implications in the impairment of gastric ulcer healing. Dig Dis Sci. 2004, 49 (5): 778-786.
104. Brakenhielm E, Cao R, Cao Y. Suppression of angiogenesis, tumor growth, and wound healing by resveratrol, a natural compound in red wine and grapes. FASEB J. 2001, 15 (10): 1798-1800.
105. Kim MH. Flavonoids inhibit VEGF/bFGF-induced angiogenesis in vitro by inhibiting the matrix-degrading proteases. J Cell Biochem. 2003, 89 (3): 529-538.
106. Ravindranath MH, Muthugounder S, Presser N, Viswanathan S. Anticancer therapeutic potential of soy isoflavone, genistein. Adv Exp Med Biol. 2004, 546: 121-165.
107. Harris ED. A requirement for copper in angiogenesis. Nutr Rev. 2004, 62 (2): 60-64.
108. Jiang C, Kim KH, Wang Z, Lu J. Methyl selenium-induced vascular endothelial apoptosis is executed by caspases and principally mediated by p38 MAPK pathway. Nutr Cancer. 2004, 49 (2): 174-183.
109. Michaud SE, Menard C, Guy LG, Gennaro G, Rivard A. Inhibition of hypoxia-induced angiogenesis by cigarette smoke exposure: impairment of the HIF-1alpha/VEGF pathway. FASEB J. 2003, 17 (9): 1150-1152.

Chapter 2

# STUDYING IN VIVO DYNAMICS OF VASCULOGENESIS USING TIME-LAPSE COMPUTATIONAL IMAGING

Evan A. Zamir[1], Paul A. Rupp[2], Charles D. Little[1]
*Department of Anatomy and Cell Biology, University of Kansas Medical Center[1], Stowers Institute for Medical Research[2]*

## 1. INTRODUCTION

The cardiovascular system is necessarily the first functional organ system in the vertebrate embryo due to metabolic requirements that cannot be met by diffusion alone. Primary vasculogenesis, or the *de novo* formation of blood vessels in the embryo, begins with the coalescence of individual mesodermally-derived endothelial cells into an intricate polygonal network (Figure 2-1) — these events occur prior to the onset of circulation. Vasculogenesis proceeds in a cranial to caudal and dorsal to ventral fashion as the embryonic axis elongates[1-3]. It is important to note that vasculogenesis is not restricted to embryogenesis proper, and indeed, similar nascent processes have recently been found to play a significant role in adult vascularization, such as in tumor formation, ischemia and vascular repair[4,5].

For convenience, the sequence of primary vasculogenesis can be delineated into several distinct stages[6,7]: (1) Endothelial precursors (angioblasts) undergo an epithelial to mesenchymal transition and migrate ventrally from the splanchnic mesoderm to the splanchnopleural extracellular matrix (ECM). (2) In a process that is still not fully understood, angioblasts then differentiate into endothelial cells. There is evidence that the differentiation is mediated by inductive growth factors, possibly FGF-2, secreted from the ventrally adjacent endoderm[8,9]. (3) Isolated clusters of endothelial cells form a planar



*R. Forough (ed.), New Frontiers in Angiogenesis,* 31–44.

polygonal pattern of endothelial chords that initially do not contain lumens. There is now strong evidence that the Hedgehog signaling pathway is involved in the process of tubulogenesis — the stage when chords become tubes[10]. (4) Finally, the primary vascular network expands by further recruitment of endothelial cells, protrusive extensions from existing branches, and vascular fusion[11,12]. During primary vasculogenesis, the newly formed vessels are free of smooth muscle cells, pericytes, or other associated cells.

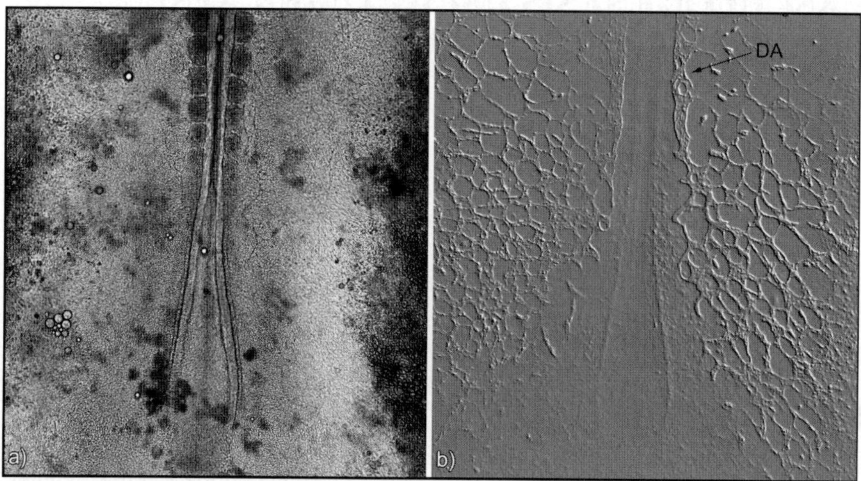


*Figure 2-1.* Images of caudal portion of 6-somite quail embryo showing in a) brightfield and b) EC vascular network, labeled by mAB to QH1 (embossed to enhance detail). Note the asymmetry in vascular patterning between the two lateral sides of the embryo, indicating the emergent nature of the phenomenon. The dorsal aortae (DA) lateral to the midline form during these stages, as EC ingress from the lateral plate mesoderm.

Work in the past decade has led to the identification of several molecular factors required for normal vascular development in avian[8,13,14], mouse[15-18], and more recently, zebrafish embryos[19-21]. The biophysical mechanisms that drive endothelial cell network formation in each species are poorly understood. Moreover, it appears that the "emergent" nature of vascular pattern formation in the warm-blooded embryo (Figure 2-1)[14], likely due to epigenetic phenomena not directly encoded in the genome, may be fundamentally different from the pre-programmed or "hard-wired" vascular network formation seen in lower vertebrates, such as zebrafish[22,23].

Two general classes of mathematical models for vasculogenesis have been proposed — one is based on the original mechanochemical theory of Murray and Oster[24-27], and the other on percolation theory[28-30]. The most important mechanistic difference between these models is that the first assumes mechanical forces (via cell-ECM traction) are primarily responsible for vascular patterning, while the latter is based on underlying chemoattractant (i.e. signaling) forces arising from the endothelial cells themselves, although the experimental evidence for this hypothesis is somewhat tenuous. Perhaps, not surprisingly, both of these models reproduce in great detail the prototypical polygonal pattern seen in vasculogenesis models *in vitro*[17,31], and to some extent, *in vivo*. Formulating theoretical constructs for these models, based on the results for controlled *in vitro* experimental systems, is clearly useful; however, there have been few, if any, attempts to verify the validity of these models in living embryos. It should be noted that the two models do not have to be mutually exclusive, and one or the other may be more appropriate for describing certain stages involved in vasculogenesis.

## 2. AVIAN TIME-LAPSE COMPUTATIONAL MICROSCOPY

Recently, a novel method for multi-scale, computational, time-lapse microscopy was developed for imaging the entire avian embryo with cellular resolution during early stages of development[11,22,34-36]. It is now possible to readily observe events taking place with length scales ranging from the order of 1 mm (e.g. macroscopic or large-scale morphogenetic tissue deformations) down to 1 μm (cell migration). This is accomplished by rapidly acquiring digital images with widefield 10x or 20x objectives, typically for up to 12 hours at multiple x-y locations and focal planes ("z-stacks"). The raw images are then processed using sophisticated, custom-written image processing software that automatically merges and collapses the thousands of image files (or tiles) and several gigabytes of data into a single 2-D time-lapse sequence or "movie"[36]. In addition to capturing differential interference contrast (DIC) or brightfield images, up to three fluorescent illumination channels can be captured during a single experiment, thus allowing the motion of both cells and matrix constituents to be tracked simultaneously using fluorescently-conjugated antibodies or transfected GFP-constructs[11,34]. The temporal resolution of these experiments is on the order of 1-10 minutes/frame, with the frame rate dependent on the number of embryos selected, illumination modes (i.e. fluorochromes), fields, exposure times, and camera resolution setting — all of which are computer-controlled parameters determined by the investigator. During an experiment, embryos develop *ex vivo* in a custom-built, temperature- and humidity-controlled culture chamber.

## 2.1 Time-Lapse Studies of Quail Vascular Development

The Japanese quail (*Coturnix coturnix japonica*) embryo provides several technical advantages for studying vertebrate vessel formation, and vasculogenesis in particular[22]: (1) It is relatively easy to obtain embryos for studying any stage of vascular development. (2) There exists a monoclonal antibody (QH1) that specifically recognizes and strongly binds a quail endothelial cell-specific surface antigen. (3) Vasculogenesis occurs essentially in a single plane. (4) Finally, vasculogenesis proceeds in a head-to-tail fashion, so that it is possible to observe the various stages of vasculogenesis within a single embryo.

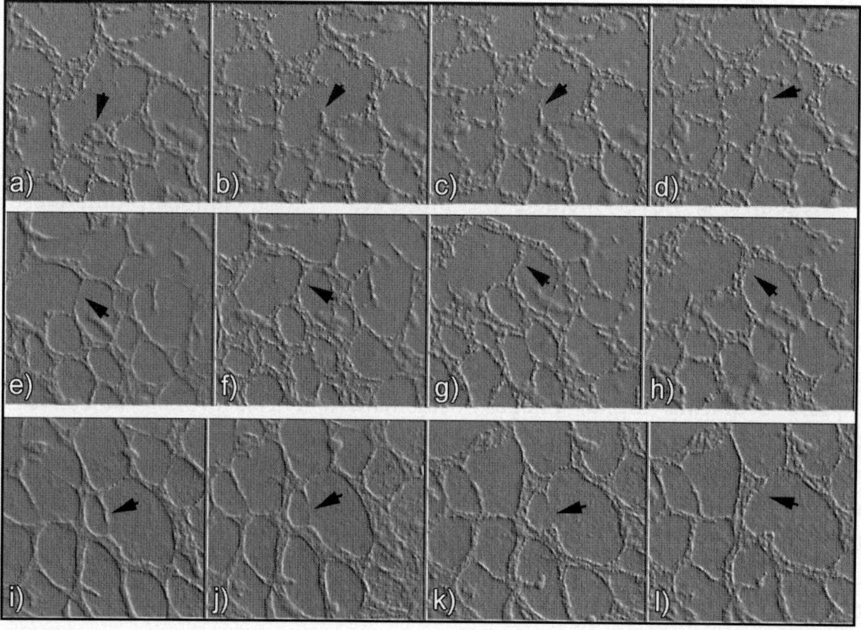


*Figure 2-2.* Time-lapse sequences demonstrating types of EC protrusive (i.e. branching) behavior during vascular network formation: (a-d) formation, (e-h) thickening, and (i-l) rupturing.

Using our computational microscopy approach, it was shown that the formation of the primordial vascular bed involves a hierarchy of movements, ranging from single-cell motility to tissue-level deformations[11,22]. At a relatively small length-scale, there is the single endothelial cell that can behave in a variety of ways — such as migrating along existing vascular structures, sending cellular processes (i.e. filopodia) across avascular zones to form new links in the network, or conversely, retracting extensions reducing the complexity of the network. At a larger length-scale are groups of cells that work in concert to structurally rearrange the vascular polygons, for example, resulting in the formation of larger caliber vessels by vascular fusion and the addition of new cells to the endothelial network. Moving up the length-scale hierarchy once again brings us to the idea of "vascular drift", a process whereby the entire vascular plexus slowly shifts or translates medially towards the axis. Finally, we arrive at the scale of tissue-level or macroscopic deformations that passively convect endothelial cells as the embryonic plate folds and elongates. In addition to these observations, others have been recorded as well and will be discussed below.

## 2.2 Cell-Level Vascular Rearrangements

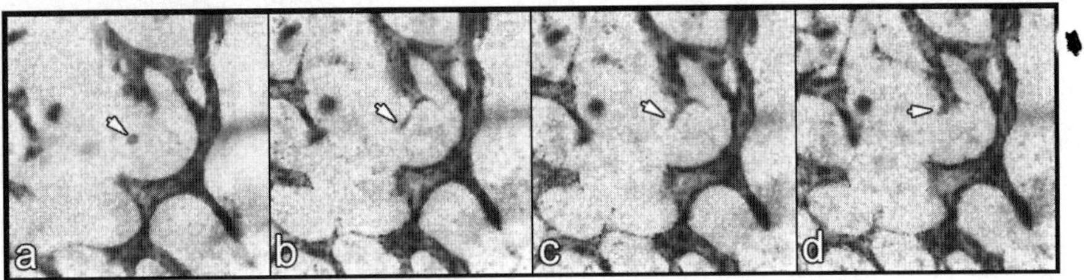


*Figure 2-3.* Isolated EC cluster (arrowhead), perhaps, a single cell, is shown joining up with the pre-existing vascular network. Without a nuclear label, it is somewhat difficult to determine the number of cells in a cluster, which have been shown to be densely packed7. Although this type of event presumably occurs quite often, it is actually quite difficult to capture, due to the limitations of time-lapse and the pulse-labeling method (see text).

Some of the characteristic morphologic behaviors of the developing vascular network mentioned above are shown in Figure 2-2, including (1) the formation of new branches (Figure 2-2, a-d); (2) the "thickening" of existing branches by additional endothelial cells (Figure 2-2, e-h), which is presumably accompanied by a proportional increase in lumen width; and (3) the regression or breaking of existing branches (Figure 2-2, i-l). This last phenomenon is particularly interesting, because it is unclear whether branches break simply due to mechanical stresses or regress due to autonomous endothelial cell behavior. We have observed that retracting branches tend to be extremely thin, possibly containing only one or two cells. Conversely, branches in the process of connecting to existing polygons are always observed to be thicker. It is therefore possible that a newly protruding or open-ended branch is under less mechanical stress than a vascular cord that is connected to the network at both ends. If the vascular network exhibits a resting tension or residual stress, it would make sense that a newly formed branch would generate some increased force to re-insert itself into the network and maintain some sort of mechanical or tensional homeostasis[37,38].

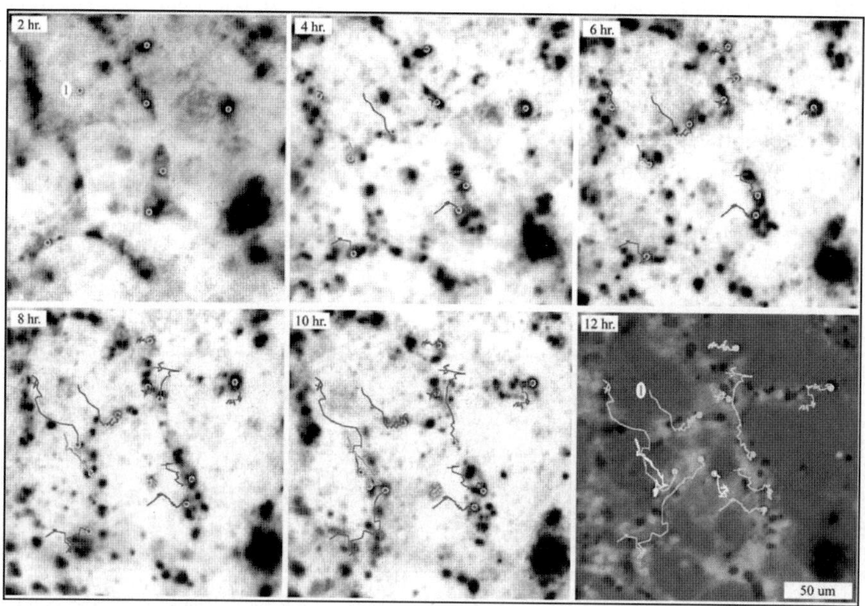


*Figure 2-4.* Trajectories of individual primordial endothelial cells, relative to the surrounding vascular structures. Each panel is a 200 μm x 200 μm area with the embryonic midline on the right, encompassing a 10 hour period of normal development. Circles highlighted by yellow indicate the current position of representative cells; lines show their trajectory up to the corresponding time point. Red and green color is used to distinguish medially and laterally oriented motion, respectively. Note, even with the subtraction of the medial vascular drift, there is a heavily favored medial-to-lateral cell migration. In the final panel (12 hr.), the 33 experimentally-labeled cells (black) and cell trajectories (yellow or white) overlay the post-experimentally labeled vasculature (green). Cell #1 first appears in an avascular zone and quickly migrates to a chord structure. Reproduced from Rupp et al., 2004[11].

Occasionally, isolated endothelial cell clusters or single cells are seen fusing with extant polygons (Figure 2-3 and Figure 2-4, cell #1). These isolated cells often appear in avascular zones and are believed to be newly differentiated endothelial cells (or cells newly arriving in the splanchnopleural ECM). It is difficult to capture the process of "new" isolated cells joining the vascular bed for two main reasons. First, it appears that isolated cells or cell clusters in the mesoderm integrate very rapidly into nearby polygons[11] — this presents a considerable challenge for *in vivo* time-lapse microscopy. In other words, the temporal resolution (minutes) of current time-lapse systems (including ours) may not be great enough to capture all these events. Second, the QH1-antibody used to visualize differentiated endothelial cells acts as a pulse-label; therefore, any angioblasts undergoing differentiation after the labeling period (at the beginning of the experiment) are not marked. Thus, it has been technically difficult to observe (record) QH1-expressing cells that exist only fleetingly as observable, isolated cell clusters. Of course, this feature of *in vivo* vasculogenesis is distinct from *in vitro* studies, in which isolated cell clusters can be observed for relatively long periods of time, even days, at high spatial resolution.

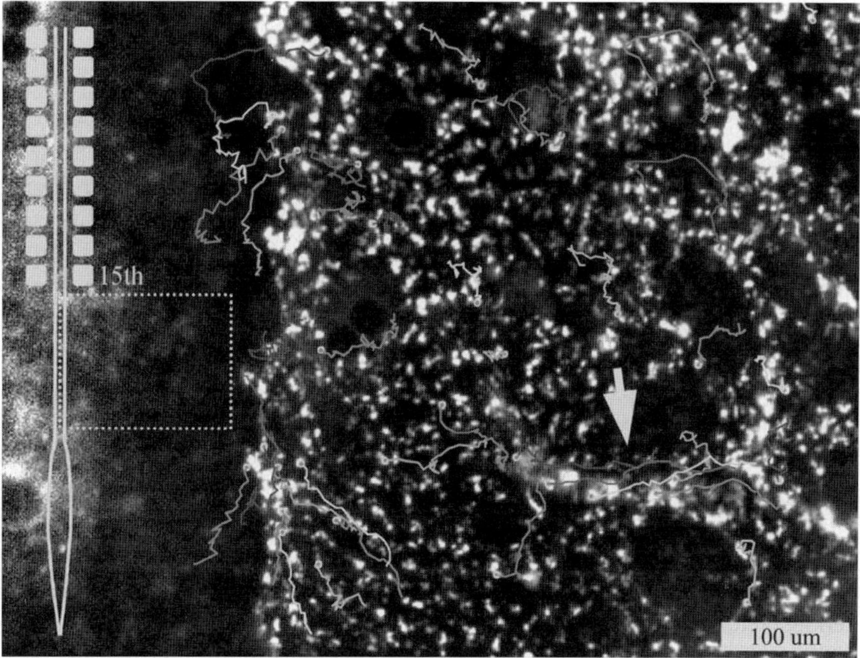


*Figure 2-5.* Trajectories of endothelial cells within the site of the future omphalomesenteric vessels. Endothelial cells were labeled with Cy3-QH1 and tracked over an 11-hour time period. Yellow highlighted circles indicate the position of individual endothelial cells at the end of the recorded period. The colored lines represent the trajectories of the corresponding endothelial cells. The arrow indicates individual cells streaming along an existing vessel toward the midline (dorsal aortae). The area analyzed is just caudal to the last formed somite (15th pair) and is depicted by the box shown in the embryonic axis diagram. Scale bar equals 100 μm. Reproduced from Rupp et al., 2003[22].

A method for labeling angioblasts would help overcome this technical hurdle, allowing the transition from angioblasts to differentiated endothelial cells to be observed more precisely *in vivo*. At present, the best angioblast-specific marker is TAL1/SCL[7]; however, there are currently no reports of live-cell studies using this gene, *in vivo*. A less direct method is to non-specifically transfect cells with a fluorescent marker, using either electroporation or retroviral techniques. In this manner, it may be possible to identify angioblasts by subsequent co-localization with QH1 after fixation. The time-lapse movie would then be played backwards, essentially following a selected endothelial cell back to its "birth". Clearly, this method requires the kind of wide-field time-lapse microscopy techniques we (and others) are developing. A current limitation of avian transfection, however, is the time needed to observe robust protein expression, often on the order of several hours.

Another phenomenon readily observed with time-lapse analysis is the formation of the great vessels, namely the dorsal aortae, sinus venosae and omphalomesenteric (yolk) vessels. The large yolk vessels form, in part, by vascular fusion of smaller nascent vessels[22]. In addition to the fusion of nascent endothelial tubes, we identified an influx of highly motile endothelial cells "streaming" in from the lateral extraembryonic regions. Some of these streaming cells populate the omphalomesenteric vessels (Figure 2-5), but most migrate medially to the site of the future dorsal aortae. Preliminary data suggest that a small number of these cells continue medially and approach the notochordal axis, but do not cross the midline. These streaming endothelial cells then appear to turn and migrate both cranially and caudally — thus beginning formation of the intersomitic and vertebral vessels.

Interestingly, the streaming endothelial cells contribute to both arteries and veins — specifically, the omphalomesenteric veins and the dorsal aortae (unpublished data). This observation raises interesting questions regarding arterio-venous fates. Studies have suggested that members of the Eph/ephrin cell surface protein families are functional determinates of artery versus venous fate determination[39-44]. Because the streaming endothelial cells we discovered originate in distant extraembryonic sites and contribute to both arteries and veins, the cells must either have extraembryonically-determined aterio-venous fates or are naive with respect to anterio-venous fate. If the latter case is true, and the streaming cells are not committed to a specific vessel fate, then it follows that primordial arteries and veins can be partially assembled by endothelial cells that may not obey the Eph/ephrin patterning rules — raising a question as to whether the Eph/ephrins are markers of arterial/venous fate or are spatial markers that delineate anatomical boundaries. Further evidence of an acquired arterial or venous fate was recently reported when Eichman and her colleagues showed that flow regulates the activation of the arterial markers ephrinB2 and neuropilin 1 in chick embryo yolk sac vessels[45]; suggesting that early endothelial cells are plastic with respect to their vessel fate.

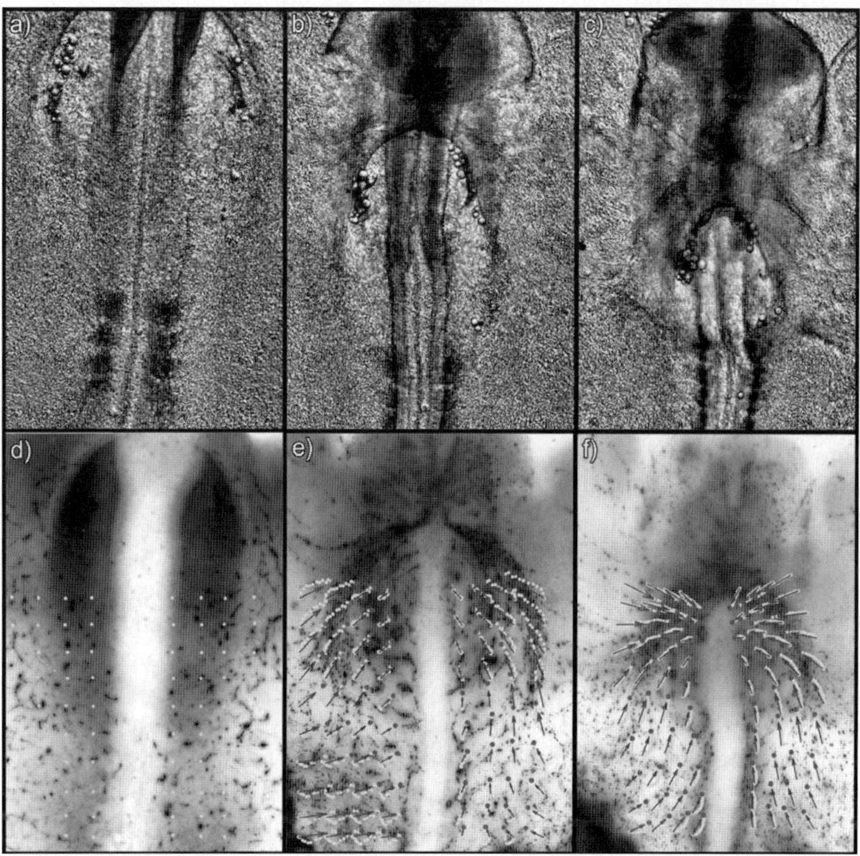


*Figure 2-6.* Development of heart and omphalomesenteric veins (a-c), shown here in a time-lapse sequence starting with a 3-somite quail embryo, exerts forces on vascular EC network (d-f), stained with QH1 antibody, causing large-scale displacements and deformations. Trajectories (yellow dots) for some arbitrary (virtual) network locations were automatically computed using a PIV method. Red dots mark the end of each trajectory, and the blue arrows represent the (normalized) displacement vectors for the tracked points during the experiment.

## 3. EFFECTS OF LARGE-SCALE MORPHOGENETIC DEFORMATIONS ON VASCULAR PATTERN FORMATION

Above we discussed some local or microscopic rearrangements in the emerging vascular pattern that are primarily due to endothelial cell migration, cell protrusion or the retraction of endothelial processes. In addition to the local effects, there are large-scale or macroscopic patterning phenomena that influence the vascular pattern. Several fundamental morphogenetic processes, including heart and gut formation and notochord regression (axial elongation), exert mechanical stresses on surrounding tissues. These tissue-level deformations passively convect cells large distances in the embryo.

Recently, a method for automated, computational "tissue fate mapping" was developed using digital particle image velocimetry (PIV), a robust image correlation technique, for quantifying large-scale tissue motion and deformation in the embryo[32]. Using PIV, it was confirmed that the vascular network caudal to the developing heart appears to be "pulled" up into the presumptive endocardium (Figure 2-6) where the endothelial tubes fuse (vascular fusion) to form the endocardium. In other words, endothelial cells in the endocardium are born outside of the heart and are translocated into the heart primarily by passive tissue convection. Further experimental study and computational modeling will be necessary to identify and characterize the origin of the biophysical forces driving this macroscopic process. This work will involve biomechanical testing methods for studying soft tissues, such as microindentation[46-48]. It will also be interesting to study whether, and how, mechanical stress in this region of the embryo affects endothelial cell motility, since there appears to be additional autonomous cell motility superposed upon the macroscopic network motion[11].

In addition to the anterior-directed motion near the heart, we have observed lateral-to-medial translocation of the vascular network in the caudal region of the embryo (Figure 2-7). This effect has been referred to as "vascular drift", a sheet-like medial translocation of the vascular plexus[11]. The empirical data demonstrating long-range motion of the primary vascular plexus correlates well with recent studies of ECM dynamics at vasculogenic stages. This motion appears to be caused by large-scale embryonic tissue forces, as demonstrated by the fact that the ECM fibrillin-2 network undergoes a similar motion during these stages (Figure 2-7, cd)[33]. These data suggest that the vascular capillary plexus and its surrounding splanchnic ECM, as exemplified by fibrillin-2, are a composite structure. We hypothesize that the forces of notochord extension, combined with heart and gut formation, essentially stretch the embryo in an anterior to posterior fashion, thus, producing some lateral compression in the vascular plexus and accounting for the apparent vascular drift phenomenon.

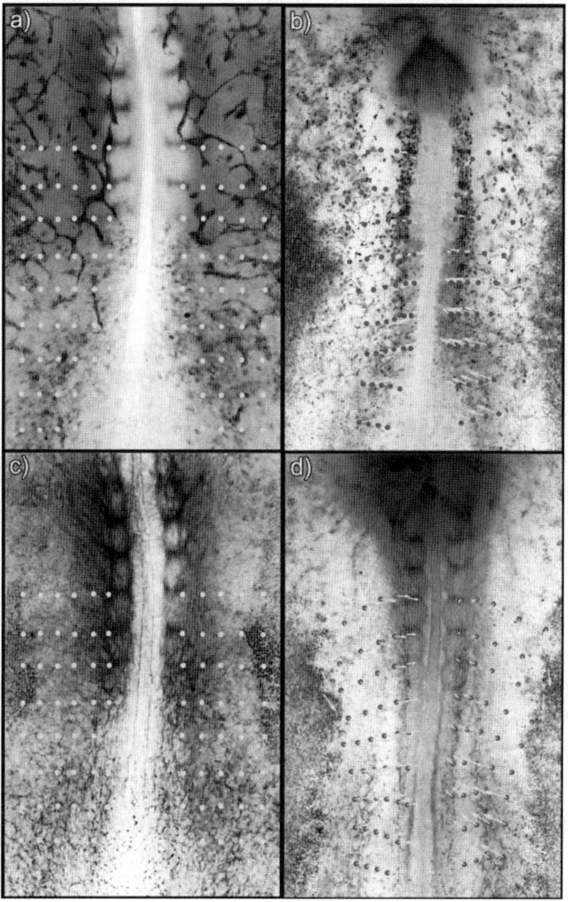


*Figure 2-7.* Comparison of tissue fate maps for vascular network (QH1-antibody, a and b) and fibrillin-2 network (JB3-antibody, c and d) in the caudal region of the same embryo after approximately 10 h of incubation.

## 4. CONCLUSIONS

In summary, through the use of multi-scale resolution, computational time-lapse microscopy, we have provided a general outline of morphogenic events during primary vasculogenesis in warm-blooded animals. We hope our data demonstrate and emphasize that the formation of the vascular bed requires coordinated motion across several length scales (from μm to mm), and that cell protrusive activity, cell locomotion, and macroscopic tissue deformations all act in concert to create an emergent and unique vascular network pattern for each embryo. While identifying and understanding the requisite molecular and genetic regulatory mechanisms are crucial, we believe it is also critical to understand the physical mechanisms required for vasculogenesis. We further hope that knowledge of these epigenetic phenomena will not only shed light on primary vasculogenesis, but also may contribute to solving such difficult problems as vascularizing bioengineered tissues or preventing neo-vascularization during tumor formation.

## REFERENCES


1. Coffin, JD and Poole, TJ. Endothelial cell origin and migration in embryonic heart and cranial blood vessel development. Anat Rec. 1991, 231, 383-395.
2. Drake, CJ, Davis, LA, Walters, L, and Little, CD. Avian vasculogenesis and the distribution of collagens I, IV, laminin, and fibronectin in the heart primordia. J Exp Zool. 1990, 255, 309-322.
3. Coffin, JD and Poole, TJ. Embryonic vascular development: immunohistochemical identification of the origin and subsequent morphogenesis of the major vessel primordia in quail embryos. Development. 1988, 102, 735-748.
4. Drake, CJ. Embryonic and adult vasculogenesis. Birth Defects Res Part C Embryo Today. 2003, 69, 73-82.
5. Asahara, T, Murohara, T, Sullivan, A, Silver, M, van der, ZR, Li, T, Witzenbichler, B, Schatteman, G, and Isner, JM. Isolation of putative progenitor endothelial cells for angiogenesis. Science. 1997, 275, 964-967.
6. Drake, CJ, Hungerford, JE, and Little, CD. Morphogenesis of the first blood vessels. Ann N Y Acad Sci. 1998, 857, 155-179.
7. Drake, CJ, Brandt, SJ, Trusk, TC, and Little, CD. TAL1/SCL is expressed in endothelial progenitor cells/angioblasts and defines a dorsal-to-ventral gradient of vasculogenesis. Dev Biol. 1997, 192, 17-30.
8. Poole, TJ, Finkelstein, EB, and Cox, CM. The role of FGF and VEGF in angioblast induction and migration during vascular development. Dev Dyn. 2001, 220, 1-17.
9. Cox, CM and Poole, TJ. Angioblast differentiation is influenced by the local environment: FGF-2 induces angioblasts and patterns vessel formation in the quail embryo. Dev Dyn. 2000, 218, 371-382.
10. Vokes, SA, Yatskievych, TA, Heimark, RL, McMahon, J, McMahon, AP, Antin, PB, and Krieg, PA. Hedgehog signaling is essential for endothelial tube formation during vasculogenesis. Development. 2004, 131, 4371-4380.
11. Rupp, PA, Czirok, A, and Little, CD. {alpha}v{beta}3 integrin-dependent endothelial cell dynamics in vivo. Development. 2004, 131, 2887-2897.
12. Drake, CJ and Little, CD. VEGF and vascular fusion: implications for normal and pathological vessels. J Histochem Cytochem. 1999, 47, 1351-1356.
13. Drake, CJ, LaRue, A, Ferrara, N, and Little, CD. VEGF regulates cell behavior during vasculogenesis. Dev Biol. 2000, 224, 178-188.

14. Flamme, I, Breier, G, and Risau, W. Vascular endothelial growth factor (VEGF) and VEGF receptor 2 (flk-1) are expressed during vasculogenesis and vascular differentiation in the quail embryo. Dev Biol. 1995, 169, 699-712.
15. Drake, CJ and Fleming, PA. Vasculogenesis in the day 6.5 to 9.5 mouse embryo. Blood. 2000, 95, 1671-1679.
16. Miquerol, L, Gertsenstein, M, Harpal, K, Rossant, J, and Nagy, A. Multiple developmental roles of VEGF suggested by a LacZ-tagged allele. Dev Biol. 1999, 212, 307-322.
17. Shalaby, F, Rossant, J, Yamaguchi, TP, Gertsenstein, M, Wu, XF, Breitman, ML, and Schuh, AC. Failure of blood-island formation and vasculogenesis in Flk-1-deficient mice. Nature. 1995, 376, 62-66.
18. Millauer, B, Wizigmann-Voos, S, Schnurch, H, Martinez, R, Moller, NP, Risau, W, and Ullrich, A. High affinity VEGF binding and developmental expression suggest Flk-1 as a major regulator of vasculogenesis and angiogenesis. Cell. 1993, 72, 835-846.
19. Roman, BL and Weinstein, BM. Building the vertebrate vasculature: research is going swimmingly. Bioessays. 2000, 22, 882-893.
20. Vogel, AM and Weinstein, BM. Studying vascular development in the zebrafish. Trends Cardiovasc Med. 2000, 10, 352-360.
21. Fouquet, B, Weinstein, BM, Serluca, FC, and Fishman, MC. Vessel patterning in the embryo of the zebrafish: guidance by notochord. Dev Biol. 1997, 183, 37-48.
22. Rupp, PA, Czirok, A, and Little, CD. Novel approaches for the study of vascular assembly and morphogenesis in avian embryos. Trends Cardiovasc Med. 2003, 13, 283-288.
23. Weinstein, BM. What guides early embryonic blood vessel formation? Dev Dyn. 1999, 215, 2-11.
24. Namy, P, Ohayon, J, and Tracqui, P. Critical conditions for pattern formation and in vitro tubulogenesis driven by cellular traction fields. J Theor Biol. 2004, 227, 103-120.
25. Murray, JD. On the mechanochemical theory of biological pattern formation with application to vasculogenesis. C R Biol. 2003, 326, 239-252.
26. Manoussaki, D, Lubkin, SR, Vernon, RB, and Murray, JD. A mechanical model for the formation of vascular networks in vitro. Acta Biotheor. 1996, 44, 271-282.
27. Murray, J, Oster, G, and Harris, A. A mechanical model for mesenchymal morphogenesis. J Math Bio. 1983, 17, 125-9.
28. Ambrosi, D, Gamba, A, and Serini, G. Cell directional and chemotaxis in vascular morphogenesis. Bull Math Biol. 2004, 66, 1851-1873.
29. Serini, G, Ambrosi, D, Giraudo, E, Gamba, A, Preziosi, L, and Bussolino, F. Modeling the early stages of vascular network assembly. EMBO J. 2003, 22, 1771-1779.
30. Gamba, A, Ambrosi, D, Coniglio, A, de Candia, A, Di Talia, S, Giraudo, E, Serini, G, Preziosi, L, and Bussolino, F. Percolation, morphogenesis, and burgers dynamics in blood vessels formation. Phys Rev Lett. 2003, 90, 101-118.
31. Vernon, RB, Lara, SL, Drake, CJ, Iruela-Arispe, ML, Angello, JC, Little, CD, Wight, TN, and Sage, EH. Organized type I collagen influences endothelial patterns during "spontaneous angiogenesis in vitro": planar cultures as models of vascular development. In Vitro Cell Dev Biol Anim. 1995, 31, 120-131.
32. Zamir, EA, Czirok, A, Rongish, BJ, and Little, CD. A Digital Image-Based Method for Computational Tissue Fate Mapping during Early Avian Morphogenesis. Ann Biomed Eng. 2005, 33, 854-865.
33. Czirok, A, Rongish, BJ, and Little, CD. Extracellular matrix dynamics during vertebrate axis formation. Dev Biol. 2004, 268, 111-122.

34. Filla, MB, Czirok, A, Zamir, EA, Little, CD, Cheuvront, TJ, and Rongish, BJ. Dynamic imaging of cell, extracellular matrix, and tissue movements during avian vertebral axis patterning. Birth Defects Res Part C Embryo Today. 2004, 72, 267-276.
35. Rupp, PA, Rongish, BJ, Czirok, A, and Little, CD. Culturing of avian embryos for time-lapse imaging. Biotechniques. 2003, 34, 274-278.
36. Czirok, A, Rupp, PA, Rongish, BJ, and Little, CD. Multi-field 3D scanning light microscopy of early embryogenesis. J Microsc. 2002, 206, 209-217.
37. Mizutani, T, Haga, H, and Kawabata, K. Cellular stiffness response to external deformation: tensional homeostasis in a single fibroblast. Cell Motil Cytoskeleton. 2004, 59, 242-248.
38. Brown, RA, Prajapati, R, McGrouther, DA, Yannas, IV, and Eastwood, M. Tensional homeostasis in dermal fibroblasts: mechanical responses to mechanical loading in three-dimensional substrates. J Cell Physiol. 1998, 175, 323-332.
39. Lawson, ND, Scheer, N, Pham, VN, Kim, CH, Chitnis, AB, Campos-Ortega, JA, and Weinstein, BM. Notch signaling is required for arterial-venous differentiation during embryonic vascular development. Development. 2001, 128, 3675-3683.
40. Moyon, D, Pardanaud, L, Yuan, L, Breant, C, and Eichmann, A. Selective expression of angiopoietin 1 and 2 in mesenchymal cells surrounding veins and arteries of the avian embryo. Mech Dev. 2001, 106, 133-136.
41. Othman-Hassan, K, Patel, K, Papoutsi, M, Rodriguez-Niedenfuhr, M, Christ, B, and Wilting, J. Arterial identity of endothelial cells is controlled by local cues. Dev Biol. 2001, 237, 398-409.
42. Adams, RH, Wilkinson, GA, Weiss, C, Diella, F, Gale, NW, Deutsch, U, Risau, W, and Klein, R. Roles of ephrinB ligands and EphB receptors in cardiovascular development: demarcation of arterial/venous domains, vascular morphogenesis, and sprouting angiogenesis. Genes Dev. 1999, 13, 295-306.
43. Gerety, SS, Wang, HU, Chen, ZF, and Anderson, DJ. Symmetrical mutant phenotypes of the receptor EphB4 and its specific transmembrane ligand ephrin-B2 in cardiovascular development. Mol Cell. 1999, 4, 403-414.
44. Wang, HU, Chen, ZF, and Anderson, DJ. Molecular distinction and angiogenic interaction between embryonic arteries and veins revealed by ephrin-B2 and its receptor Eph-B4. Cell. 1998, 93, 741-753.
45. le Noble, F, Moyon, D, Pardanaud, L, Yuan, L, Djonov, V, Matthijsen, R, Breant, C, Fleury, V, and Eichmann, A. Flow regulates arterial-venous differentiation in the chick embryo yolk sac. Development. 2004, 131, 361-375.
46. Zamir, EA, Srinivasan, V, Perucchio, R, and Taber, LA. Mechanical asymmetry in the embryonic chick heart during looping. Ann Biomed Eng. 2003, 31, 1327-1336.
47. Zamir, EA and Taber, LA. Material properties and residual stress in the stage 12 chick heart during cardiac looping. J Biomech Eng. 2004, 126, 823-830.
48. Zamir, EA and Taber, LA. On the effects of residual stress in microindentation tests of soft tissue structures. J Biomech Eng, 126. 2004, 276-283.

Chapter 3

# THE OLD AND NEW OF BONE MARROW-DERIVED ENDOTHELIAL CELL PRECURSORS

Gina C. Schatteman[1], Ola Awad[2], and Martine Dunnwald[3]
*Departments of Exercise Science[1] Anatomy and Cell Biology[2], and Dermatology[3], University of Iowa*

## 1. SCOPE

Clinical trials involving bone marrow-derived (BMD) cells to treat non-hematopoietic disorders are underway around the world, despite the fact that we know little about their biology in non-hematopoietic tissues. We are, however, beginning to make some progress, in part by using hematopoietic stem cell biology as a guide. One particular area of interest is in BMD endothelial cell precursors. This chapter will review some of the history leading to our current interest in these cells, and discuss the difficulties in defining an 'endothelial progenitor cell'. It will also attempt to make sense of the conflicting data regarding their physiological significance. Finally, our current understanding of how BMD cells may be used therapeutically to modulate vascular growth and tissue repair will be discussed. Though cells other than BMD cells may be able to differentiate or trans-differentiate into endothelial cells, this review will focus exclusively on cells of the bone marrow. As far as we know, all of the concepts apply equally well to these other cell types.

## 2. INTRODUCTION

Therapeutic modalities for many conditions have been limited to maximally preserving existing tissue. For example, once a hemorrhage is



*R. Forough (ed.), New Frontiers in Angiogenesis,* 45–78.

stopped or a clot 'busted' and flow to the ischemic neural tissue restored, there is little else to do after a stroke but wait and hope. Similarly, after a myocardial infarct, if coronary arteries are unblocked or bypassed, the heart simply has to heal itself. While our abilities to preserve tissue have improved dramatically over the years, it is only in the last decade that we have begun to think seriously about not just preserving, but actually repairing tissues that were heretofore considered irreparable. This revolution in our thinking has been made possible in part by our enhanced abilities to manipulate embryonic stem cells, and the recognition that adult stem and progenitor cells are present in many different tissues.

The idea of an adult stem cell *per se* is not new, with hematopoietic stem cells and satellite cells reported as early as 1961.[1,2] Hematopoietic stem cells were ultimately shown to differentiate into all of the mature circulating blood cells: erythrocytes, granulocytes, and cells of the myeloid and lymphoid lineages,[3,4] while muscle satellite cells appeared to differentiate into muscle only.[5] Since then a number of other adult somatic stem cells have been identified. Adult mesenchymal stem cells located in the bone marrow stroma can be induced to differentiate into a variety of mesodermal lineage cells, although the full extent of their multipotency has taken decades to appreciate.[6-10] In the intestine, multipotent intestinal crypt stem cells provide all four cell types of the small intestine (paneth, enteroendocrine, goblet, and intestinal epithelial cells),[11,12] and epidermal and hair follicle stem cells supply the skin.[13-15] What is new is our recognition that stem cells may be present in all tissues, and that stem cells that were thought to have restricted lineage potentials are actually multipotent, and may be pluripotent.

Though a number of earlier reports hinted at broader potentials of adult stem cells, the year 1997 marked the beginning of widespread change in attitudes toward adult stem cells. In that year we reported that hematopoietic cells differentiate into ECs *in vitro* and *in vivo*. Moreover, the data suggested that hemangioblasts, a common hematopoietic and EC precursor, might be present in the blood[16]. Shortly thereafter other adult stem and progenitor cells were identified, and expanded potentials of known stem cells were described. Among the most dramatic of these findings were those describing an expanded potential for neural stem cells,[17-19] and those showing that mesenchymal stem cells are capable of giving rise to tissues of non-mesodermal origin.[10] Additional data suggested that hematopoietic cells may be the source of tissue specific stem cells such as satellite cells of skeletal muscle and liver.[20-24]

Old beliefs die hard, and there was significant resistance to the idea that hematopoietic cells might differentiate into non-hematopoietic cell types. Some of this skepticism was justified, but as evidence accumulated, this view became less and less tenable. Instead, another concern came to the fore.

If one accepts the idea that hematopoietic cells can differentiate into non-hematopoietic cells, is there a pluripotent stem or progenitor cell, or do many different progenitor populations in the bone marrow give rise to a single or a few non-hematopoietic cell types? In 2001 it was demonstrated that single bone marrow cells are capable of multi-organ multi-lineage engraftment.[25] This ultimately led to widespread acceptance of the idea that a subset of bone marrow cells are multipotent progenitors with respect to both hematopoietic and non-hematopoietic cell types. Of course, this does not rule out the possibility that different subsets of bone marrow cells with more restricted potentials also may exist and this remains an open question.

Understanding the difference between stem and progenitor cells is critical to understanding the role of bone marrow cells in non-hematopoietic tissues. Both stem and progenitor cells give rise to one or more mature cell types with characteristic morphologies and specialized functions. What distinguishes the two is the capacity of stem cells to self-renew. That is, stem cells can make identical copies of themselves, essentially for the life of the organism (although, stem cells may become dysfunctional or depleted in the aged or diseased animal).[26-28] In addition, stem cells generally divide relatively infrequently, even in tissues such as the blood and skin, which are constantly and rapidly replacing themselves. It is the task of the progenitors, their partially differentiated progeny, to proliferate extensively so as to provide replacement cells for the tissue.[29-34] True stem cells are though to be rare; hematopoietic stem cells probably represent less than 0.01% of bone marrow cells.[35]

To conclude that a cell is a stem cell, one must demonstrate long-term self-renewal. This is difficult to prove *in vivo*, though it has been demonstrated for bone marrow. The bone marrow of a recipient mouse is destroyed and subsequently reconstituted from a single donor mouse bone marrow cell.[36] Once restitution of the bone marrow is complete, the recipient's bone marrow is then transplanted to another recipient. All hematopoietic lineages can be reconstituted in the serially transplanted mice demonstrating self-renewal of stem cells.[37,38] Using this strategy Krause and colleagues demonstrated that the BMD cells capable of multi-organ multi-lineage engraftment were true stem cells.[25]

## 3. ADULT HEMANGIOBLASTS

Hemangioblasts are the embryological precursors from which blood and ECs arise. Proof of their existence was made possible by the discovery and cloning of vascular endothelial growth factor (VEGF) and its receptors VEGFR1 and VEGFR2. In one study, a single VEGFR2$^+$ cell from the chick gastrula was induced to give rise to both hematopoietic and EC colonies *in vitro*.[39] In a second study, VEGFR2$^+$ hemangioblasts were identified in early mouse inner cell mass-derived embryoid bodies.[40] While it was long accepted (without proof) that hemangioblasts were present in the embryo, it was not generally accepted that they remained extant in the adult.

The existence of adult hemangioblasts was considered unlikely because the only mechanism of blood vessel formation that had been observed in the adult was angiogenesis. In this process, ECs proliferate and migrate leading to capillary sprouting from pre-existing vessels.[41,42] Although angiogenesis also occurs in the embryo, many vessels form via vasculogenesis. During vasculogenesis, angioblasts, i.e., EC precursors derived from hemangioblasts, coalesce into tubes that ultimately form large vessels or vascular plexi.[43] Thus, if there were no angioblasts, there could be no hemangioblasts. Grant and colleagues laid that notion to rest, however, by showing that a single adult Sca-1$^+$c-kit$^+$lin$^-$ hematopoietic cell can generate both ECs and all hematopoietic cell lineages in the mouse.[44] Their findings were confirmed by Bailey et al who showed that single Sca-1$^+$ c-kit$^+$ Lin$^-$ cells reconstitute the bone marrow and contribute to the endothelium in irradiated mice.[45]

It is notable that Wagers and colleagues failed to see hemangioblast activity when a single Sca-1$^+$ c-kit$^+$ Thy1.1$^{lo}$ Lin$^-$ bone marrow cell was used to reconstitute the bone marrow.[46] The source of the discrepancy is unclear, but the most likely reason is that both Grant and Bailey intentionally damaged tissue in order to induce neovascularization.[44,45] In the absence of injury there is little vascular growth in the adult. Further, tissue damage or hypoxia may be needed to recruit, activate, or direct circulating cells into the EC lineage. It is also possible that since the tested cells were a subpopulation of those used in the other two studies, the hemangioblasts were discarded. Additionally, differences in isolation procedures and cell handling may have driven the Sca-1$^+$ c-kit$^+$ Thy1.1$^{lo}$ Lin$^-$ cell into a hematopoietic pathway. Finally, differences in the transplantation procedure and choice of accessory cells may have modulated cell potential.

## 4. CIRCULATING ENDOTHELIAL CELLS

The Grant studied proved that hemangioblasts are present in the bone marrow, but it did not address the question of whether BMD hemangioblasts or their more differentiated progeny are present in the blood. Evidence for that came from other sources. In 1994, Scott and collaborators suspended polytetrafluoroethylene pledgets in the aortas of dogs. When the pledgets were removed 55 days later, they were covered with ECs, smooth muscle cells, macrophages, monocytes, and capillary-like structures.[47] The authors concluded "The origin of the cells identified is speculative but they appear to have been derived from circulating cells, possibly stem cells, which are capable of differentiation because the pledgets on which the cells were identified were isolated from aortic wall endothelium and peri-vascular capillaries."

In hindsight, it is difficult to understand why this should have been considered a radical suggestion and why the possibility of adult angioblasts or hemangioblasts was ignored (or denied) for so long. There were no compelling data to rule out their existence, and reports had been appearing in the scientific literature since at least 1932 suggesting that circulating cells give rise to ECs. In that year, Hueper and Russel found capillary-like formations in cultures of leukocytes.[48] The following year Parker reported the development of organized vessels in cultures of blood cells, and in 1950 formation of blood vessels by cultured bone marrow cells from adult chickens was described.[49,50] In addition, two studies of thrombi concluded that cellular components of the thrombus, including various types of mesenchymal cells and ECs are derived from blood cells, not from cells of the vessel wall.[51,52] Moreover, evidence for circulating endothelializing cells of the type Scott described appeared almost as soon as the use of synthetic arterial grafts began, when it seemed that grafts were being endothelialized by circulating cells.[47,53-55] The phenomenon, dubbed 'fallout' endothelialization, occurred even when in-growth from the anastomoses and the vaso vasorum was prevented.[55,56] Further, when bone marrow cells were infiltrated into synthetic vascular grafts, endothelialization occurred more rapidly than in uninfiltrated controls.[57]

Though hemangioblasts are present in the bone marrow, cells responsible for fallout endothelialization and other vascularization phenomena could be circulating ECs sloughed from the vessel wall. However, early studies suggested that there are typically less than ten circulating ECs per milliliter of blood.[58-60] And in a survey of 21 manuscripts published since 1992 using more sophisticated technology, only three reported circulating EC concentrations of more than 20 per ml.[61-63] The number of circulating ECs does increase in conditions associated with EC dysfunction and damage such as sickle cell anemia and smoking, and in inflammatory diseases such as lupus erythematosus,[59,64-73] and it has been suggested that these may be a surrogate marker of vascular disease and tumor growth potential.[74-76] The critical question is, are the circulating ECs viable? One study found that more than 60% of circulating ECs are apoptotic under normal conditions, and another, that the cells do not survive in cell culture.[71] Still others have found <10% of circulating ECs were apoptotic in angina or myocardial infarct patients, but they also reported less than one circulating EC per milliliter of blood in these patients.[69] On the other hand, when CD146 expressing circulating cells were cultured, the cells proliferated rapidly.[77] So, one cannot rule out the possibility that circulating ECs contribute to vascular growth or fallout endothelialization.

It would seem a simple task to distinguish between hematopoietic cells and ECs, however, there is huge overlap in antigens expressed by primitive hematopoietic cells and ECs. Fortunately, three key antigens have been identified that have helped distinguish between them. CD146 (S-endo, P1H12 antigen) appears to be expressed almost ubiquitously of ECs but is not found on hematopoietic cells.[60,78,79] In contrast, CD133 (AC133) is expressed only on primitive hematopoietic cells, and probably on some or all hemangioblasts, but has not been found on ECs.[80,81] CD45 is expressed by most hematopoietic cells, but is not detected on ECs.[82,83]

The search for putative BMD circulating hemangioblasts or EC precursors began with CD34. (When these studies began it was not known that VEGFR2 was expressed on embryonic hemangioblasts.) CD34 is expressed in blood islands, early embryonic ECs, and activated adult ECs; CD34 antigenicity is routinely used to enrich for human hematopoietic stem and primitive progenitor cells.[84-88] By analogy, circulating adult EC precursors might express CD34. Accordingly, $CD34^+$ cells were isolated from fresh peripheral blood mononuclear cells (PBMCs). When cultured on fibronectin, the $CD34^+$, but not $CD34^-$ PBMCs, survived and differentiated into endothelial-like cells. After several days in culture, the cells began to express CD31 (PECAM) and tie-2 protein[89-92]; take up acetylated low-density lipoprotein (acLDL)[93,94]; bind Ulex lectin[95]; and form blood island and tube-like structure.[16] The cells also made endothelial nitric oxide synthase (eNOS) mRNA[96] and released NO in response to acetylcholine and VEGF.[16] Freshly isolated $CD34^+$ PBMCs exhibited none of these properties and did not express CD146, indicating that the cultured cells are not circulating ECs, but rather a true adult precursor population.[16]

Another group cultured circulating cells from sex-mismatched transplant patients specifically to determine if ECs derived from the blood were sloughed ECs or new BMD cells.[77] When $CD146^+$ circulating cells were placed in culture, recipient-derived cells, presumably circulating ECs proliferated rapidly, but they expanded only ~20-fold. As these cells died, a less abundant more slowly growing donor-derived (i.e., BMD) subset of cells replaced them, and these cells could be expanded more than 1000-fold. The expanded cells expressed many endothelial cell antigens. The fact that expression of a marker of mature ECs was used to select the circulating cells in these experiments probably accounts for the large proportion of circulating ECs in the cultures. Consistent with other reports that $CD146^+$ cells are rare in the blood, we do not detect CD146 on freshly plated circulating cells, but 14% of cells express it after two weeks in culture.[97] Similarly, in studies on circulating cells from bone marrow transplant patients, after pre-plating to remove adherent cells, the cultured cells differentiated into EC that were exclusively BMD and did not express CD146.[98] Thus, it appears that either the late outgrowth cells represent an extremely rare population of $CD146^+$ EC precursor or they are $CD146^-$ contaminants.

That BMD angioblasts are present in the blood was convincingly demonstrated when Sauvage and colleagues extended their earlier 'fallout endothelialization' studies.[99] First, the bone marrow of dogs was replaced with genetically distinct bone marrow. Next, a Dacron aortic graft, impervious to ingrowth from perigraft tissue and the vaso vasorum was implanted. Analysis of the graft weeks later showed that it had been re-endothelialized by cells derived from the transplanted bone marrow. In the same study, they found that $CD34^+$ cells from human bone marrow, umbilical cord, fetal liver, and mobilized peripheral blood all differentiate into endothelial-like cells that internalize acetylated low density lipoprotein and express vWF and VEGFR2 in culture.

*In vivo* proof of BMD endothelial cells in humans has come through sex-mismatched transplant patients. Chronic myelogenous leukemia patients carrying the BCR/ABL fusion gene received bone marrow transplants from non-leukemic donors. Both BCR/ABL positive and donor-derived EC were found in the patient's myocardial endothelium.[100,101] Similarly, patients that received sex mismatched bone marrow transplants were found to have male/female chimeric coronary endothelium.[102] Finally, engraftment of 2 x $10^5$ $CD34^+$ umbilical cord blood cells into NOD/scid mice resulted in bone marrow reconstitution and BMD retinal EC integration after induction of retinal hypoxia.[103]

# 5. CIRCULATING ENDOTHELIAL CELL PRECURSORS

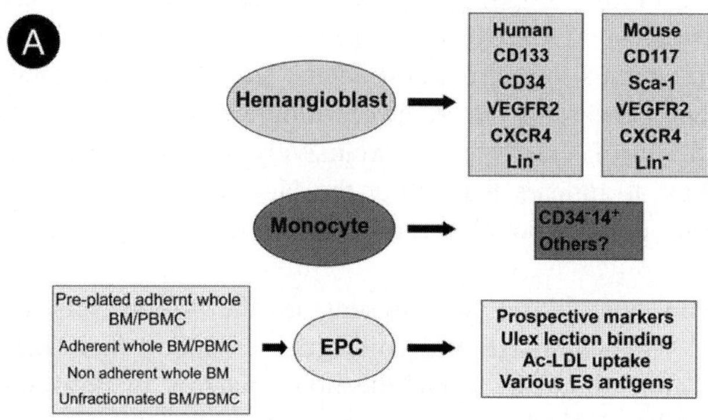


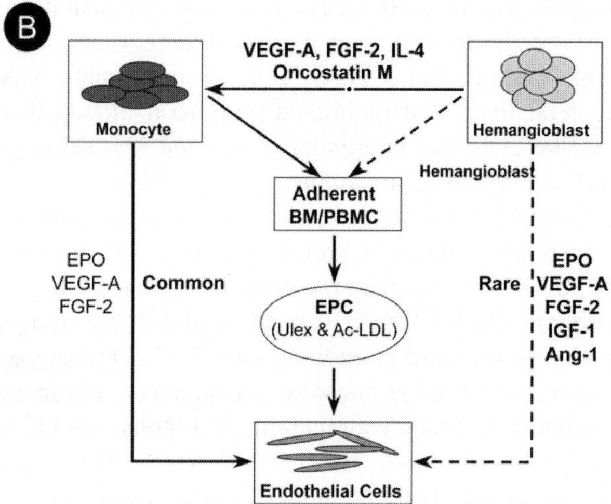


*Figure 3-1.* Classes of Bone Marrow-Derived Endothelial Cell Precursors. A) With the exception of CXCR4, the antigens listed for hemangioblasts have been used singly or in combination to select the cells from the blood or bone marrow. CXCR4 has only been used in combination with one or more of the other antigens. EPCs can be identified only prospectively, after they have taken on an endothelial cell phenotype. Binding of Ulex lectin and uptake of acetylated low-density lipoprotein are most commonly used to identify the cells. B) It is unlikely that there is a single differentiation pathway of bone marrow cells into endothelial cells. However, most, if not all, endothelial cell precursors are probably ultimately derived from some type of hemangioblast precursor, which differentiates into an endothelial cell, principally via a monocytic intermediary. The majority of adherent hematopoietic cells that differentiate into EPCs are most likely derived from a monocytic precursor as well.

These data show that two distinct populations of cells are present in the blood; endothelial-like cells derived from a non-bone marrow source, and immature BMD cells that can differentiate into ECs *in vitro* and *in vivo*. In the human, CD133 is helpful in identifying cells of bone marrow origin and CD146 marks circulating ECs, but identifying circulating EC precursors remains problematic. No consensus has been reached on the antigenic identity of these cells for at least three reasons. First, it is likely that there is no single phenotype. As in the hematopoietic system where some hematopoietic stem cells are CD34$^-$ but then turn on CD34 before producing CD34$^-$ multi-lineage progeny, the cells may well be plastic.[104,105] In addition, there simply may be many different cells that can serve the same function depending upon the environment in which they find themselves. Again, this seems to be the case for the hematopoietic system, where several mutually exclusive subsets of cells function as stem cells.[106] Second, even if we identify an antigenic profile of cells that can differentiate into ECs, *in vitro* plating and *in vivo* reconstitution efficiencies are low, so it is difficult to know if all such cells are EC precursors. Third, and most importantly, we have yet to reach a consensus of what we mean by the term 'endothelial progenitor cell'. For example, currently CD133$^+$, CD34$^+$, and VEGFR2$^+$ cells, as well as CD14$^+$ cells, adherent bone marrow cells, and non-adherent bone marrow cells have all been referred to as endothelial progenitor cells or 'EPCs'. Further within these groups, some investigators study early, and others, late outgrowth cells. Clearly these are different, albeit sometimes overlapping, cell populations with distinct characteristics.

Despite these limitations, so-called 'endothelial progenitor cells' can be roughly subdivided into three inter-related classes. (Figure 3-1) One class of EC precursors is represented by cells that are likely to be hemangioblasts. In mice this includes Sca-1$^+$c-kit$^+$lin$^-$ cells and Sca-1$^+$c-kit$^+$CD34$^-$ SP cells from the bone marrow,[44,107] and in humans, CD133$^+$, CD34$^+$, and VEGFR2$^+$ cells from the bone marrow and blood, among others.[16,108-112] Subsets of all of these populations can reconstitute the bone marrow and differentiate into ECs *in vivo*, but with one exception, we do not know if the ability to perform both of these functions resides in a single cell. Work with mesangioblasts suggests that among hemangioblast-like cells, the lack or presence of the transcription factor tal-1 distinguishes between mesangioblasts (which make mesodermal cells and EC) and hemangioblasts.[113-115] A second type of EC precursor is the monocyte or monocyte-like cell.[97,116-120] It is not known if CD14 truly defines this population, but it has been utilized widely because it is expressed on most human monocytes, but rarely, if at all, on macrophages. CD14$^+$ cells have been reported to be multipotent with respect to non-hematopoietic tissues,[118,120,121] and one study suggests that they may be multipotent with respect to hematopoiesis as well.[121] Cells that are selected prospectively represent the third class, that which is now most commonly referred to as EPCs. Typically, these are adherent total peripheral blood leukocytes or whole bone marrow cells after various lengths of time in culture. Some investigators pre-plate to remove mesenchymal stem cells and circulating or stromal ECs,[98,122,123] others do not.[124-126] The common characteristic of these cells is that they express of Ulex lectin and take up acetylated low-density lipoprotein. Depending on selection and culture conditions they also express a wide variety of EC antigens.

It has been suggested by ourselves and others, that EPCs are derived primarily from circulating monocytes, possibly $CD34^-CD14^+$ cells.[109,120] As noted, like monocytes, EPCs take up acetylated low-density lipoprotein and bind of Ulex lectin. EPCs also ultimately express CD144 (VE-cadherin), $CD34^+$, $CD31^+$, VEGFR2, and P1H12[127-129], yet after the initial four days in culture few cells express CD144, E-selectin, or CD34, but do express CD45, CD14, CD163, and CD11c.[120,130] This phenotype is more characteristic of monocyte/macrophages than EC. With time in culture, EPCs co-express monocytic and EC antigens,[130] which is consistent with studies demonstrating that monocyte derived EC initially co-express EC and monocyte antigens but lose monocytic characteristics over time.[97,118,119] Additionally, we were unable to derive ECs from a $CD34^-CD14^-$ fraction of peripheral blood.[97] Even $CD34^+$ cells may differentiate into EC through a $CD14^+$ intermediary. One study that followed their phenotype over time in culture found that $CD34^+$ cells first differentiated along the monocytic lineage, expressing $CD14^+$, with no evidence for endothelial cell differentiation. Later the cells acquired endothelial cell markers, including CD146, EC nitric oxide synthase, and von Willebrand factor.[131] Similarly, circulating $CD117^+$ cells differentiated to EC through a $CD68^+$, that is, monocytic, intermediary.[132]

A recent study suggests that expression of CD14 is not requisite for EPCs.[133] The expression of vWF was evaluated on $CD14^+$ and $CD14^-$ cells after 4 days in culture, and levels of the EC antigen were similar in $CD14^+$ and $CD14^-$ cell cultures. However, the $CD14^-$ cells were not depleted of $CD34^+$ cells, so the $CD14^-$ fraction would have contained $CD34^+$ cells that may have differentiated. In addition, the number of EPCs in the two different cultures was not mentioned, so contaminating $CD14^+$ cells could have contributed the EC-like cell population as well. $CD14^+$ and $CD14^-$ cultures did produce similar numbers of 'CFU-ECs', but the CFU-ECs were characterized only by uptake of acetylated low-density lipoprotein and binding of Ulex lectin so it is not clear if these colonies represented ECs or monocyte/macrophages. Thus it remains likely that EPCs are primarily $CD14^+$ monocyte-derived. However, due to the apparent plasticity of hematopoietic cells, it is also possible that both $CD34^-CD14^+$ and $CD34^-$

CD14⁻ cells can differentiate into EC depending on culture conditions *in vitro* and the local environment *in vivo*. CD34⁻ hematopoietic stem cells acquire CD34 expression before CD38 or lineage markers are expressed,[104,105] so a similar process could occur in CD14⁻ cells.

The confusion in the definition of EC precursors carries over to the discussion of their biology. It is not uncommon to see direct comparison of data among studies that were carried out on EPCs derived from different subsets of bone marrow or circulating cells. Additionally, even within single reports, circulating hemangioblastic precursors and adherent bone marrow or circulating cells are both referred to as EPCs. Another problem arises because EPCs are defined prospectively, i.e., after they have differentiated into ECs. Thus, they are no longer precursors, but rather EPC-derived ECs. Finally, some investigators do not distinguish between murine and human antigens. For example, while CD34 is expressed by most human hematopoietic stem cells, this is not the case for mice,[134,135] and conversely CD117 appears to be expressed on most mouse hematopoietic stem cells, but its relevance as a marker for human hematopoietic stem cells is less clear.[136-138] Yet, it has been assumed by some that CD34 is present on murine and CD117 on human EC precursors.

Although the different EC precursors share many characteristics, they are distinct, and respond differently to autocrine stimulation and disease.[109,139,140] Our failure to use precise nomenclature and to distinguish between the different populations when discussing their behaviors and functions, has undoubtedly contributed to the fact that among the clinical trials currently underway using BMD cells to induce vascular growth, each employs a different subset of mononuclear cells, introduced in a different way, at a different time point. A more precise description of cell phenotypes and characteristics in pre-clinical studies would provide better guidance for physicians wishing to apply cell based therapies in the clinic.

## 6. PHYSIOLOGICAL RELEVANCE OF BONE MARROW-DERIVED ENDOTHELIAL CELLS

It has been argued that studies on the hemangioblastic potential of the bone marrow are non-physiologic, because induction of retinal ischemia with addition of VEGF and radiation damage were required to induce integration of the cells into the endothelium.[46] However, since new blood vessel growth rarely occurs in the adult, and turnover of the quiescent endothelium is thought to be low,[58,141] little integration of bone marrow cells into the quiescent vasculature would be expected. Since it is only after injury that new ECs would be required, it seems reasonable to induce injury in order to assess integration of bone marrow into the endothelium. Consistent with this, examination of mouse tissue 16 weeks after bone marrow transplantation, revealed low but detectable levels BMD EC in the pre-existing endothelium, but significantly greater incorporation (8.3-11.2% of total EC) in the neovessels of an implanted sponge and ischemic limb.[142] Further, EC precursors are attracted to sites of injury and tissue ischemia. Injected BMD cells are preferentially localized in the myocardial infarct and peri-infarct zones of infarcted rats,[107,143] and in ischemic hindlimb models, injected EC precursors are more common in the ischemic limb than the non-ischemic limb.[16,109,144] Hindlimb ischemia, myocardial infarction, vascular trauma, pancreatic damage, and exercise all mobilize BMD EC precursors, and this mobilization correlates with increased vascularization in tissues undergoing neovascularization.[144-148] In an elegant study in which a gradient of hypoxia was created in skin wounds in mice, recruitment of BMD cells followed the gradient.[149] The greatest number of cells homed to and integrated into vessels of the most ischemic tissue. The BMD vessels comprised a significant portion of the vessels in the ischemic tissue.

Introduction of supra physiological levels of growth factors to induce vascular growth provides insight into the maximal potential of bone marrow cells to be recruited to and integrated into the vascular endothelium. In mice implanted subcutaneously with matrigel plugs impregnated with fibroblast growth factor-2, 26.5% of ECs in the neovasculature were bone marrow derived,[150] and VEGF impregnated pellets implanted into the cornea resulted in 17.7% of ECs in the neovasculature being BMD.[150] Incorporation of only 7.3% was reported in the absence of simvastatin, but drug treatment increased the percentage of BMD EC to 25.7% in another study using the same model.[151] In the mouse retinal neovascularization model described above, although not specifically quantitated, large BMD foci of neovascularization were observed.[44] Thus, bone marrow cells can make substantial contributions to the growing vasculature under certain conditions, and this has tremendous implications for understanding tumor growth and design of therapeutic modalities.

So, do BMD EC precursors potentiate tumor growth? Studies in humans showed that an average of 4.9% of EC in tumors from five patients were BMD EC.[152] BMD cells have been found in the endothelium of implanted MCA38 colon and Tg.AC line 43 tumors in mice.[153,154] BMD EC were present in the tumor vasculature of mice treated with granulocyte colony-stimulating factor (G-CSF); G-CSF is thought to mobilize EC precursors into the blood.[155] While the G-CSF had no effect on CMT colon cancer cell proliferation *in vitro*, it markedly increased tumor growth *in vivo*, and the accelerated tumor growth was associated with enhancement of neovascularization in the tumor.[155] Additional evidence for a role for BMD

EC in tumors comes from $Id1^{+/-}Id3^{-/-}$ mutant mice. These mice have a diminished angiogenic capability and are tumor resistant. BMD ECs were found in both tumors and implanted Matrigel plugs in the $Id1^{+/-}Id3^{-/-}$/wild type chimeric mice, and tumor growth was similar to that of wild type animals in mutant mice that received a wild type BM transplant.[156] Of course, if circulating EC precursors really do travel preferentially to tumors, they could be used as a gene delivery tool. To test this, BM was reconstituted with murine BM cells carrying a gene for an anti-tumor or anti-angiogenic agent. Tumor growth was significantly inhibited in the mice carrying the genetically manipulated bone marrow.[157,158]

Despite what we believe is rather compelling evidence that BMD EC are physiologically significant, it is still argued by many that they are merely an interesting biological curiosity. This is due in part to the contradictory findings reported in the literature. The primary means of examining the relevance of BMD cells in non-hematopoietic tissues has been mismatched bone marrow transplantation, and as can be concluded from the small sample of studies outlined above, a myriad of models, each with its own unique characteristics, have been employed to this end. Some of the differences in models may seem small, but these small changes may in fact account for the observed large differences in their *in vivo* physiology.

First of all, the type of cells engrafted matters. For example, $Sca\text{-}1^{+}$ bone marrow cells from mice expressing GFP under the control of Tie2 (Tie2/GFP) were used for bone marrow reconstitution.[154] $Sca\text{-}1^{+}GFP^{+}$ or $Sca\text{-}1^{+}GFP^{-}$ cells from these animals were engrafted into wild type mice. After BM reconstitution, mice were injected with tumor cells. Upon subsequent analysis, $GFP^{+}$ cells were present in the tumor endothelium of mice engrafted with $Sca\text{-}1^{+}GFP^{+}$ but not $Sca\text{-}1^{+}GFP^{-}$ cells. One interpretation is that $Sca\text{-}1^{+}Tie2^{+}$ BM cells but not $Sca\text{-}1^{+}Tie2^{-}$ BM have the potential to differentiate into EC. Of course, another possibility is that the $Sca\text{-}1^{+}GFP^{-}$ cells simply represented a population of BM cells that had permanently silenced the GFP transgene. Potential differences in engrafted populations is not limited to differences in cell surface antigens, since even the cell cycle may affect cell engraftment,[159,160] and cell cycle can be affected by the method of isolation. Further, the purity of an isolate may affect engraftment if a particular type of accessory cell specifically promotes hemangioblast engraftment.

Of course the animal model will have an impact on differentiation and incorporation of cells within a particular tissue. Even within the same species, strain differences may alter BMD EC integration. Moreover, we have found changes in EC precursor function within a single strain when mice were purchased from two different vendors. (Awad and Schatteman, unpublished) Two studies looking at the contribution of BM cells to skeletal muscle showed that the background physiology of the mouse effects BM contributions to injured tissue, and may dictate whether fusion or frank differentiation occurs.[161,162] Further, the extent of integration of BM into skeletal muscle was unique to each muscle, ranging from less than 0.01% in the tongue to more than 5% in the panniculus carnosus.[163] Given the heterogeneity of the endothelium, a similar degree of BMD EC incorporation might well be anticipated. Another consideration is that although higher levels of engraftment are observed with injury to tissue, induced injuries are not uniform with respect to hypoxia, necrosis, inflammation, and molecular changes. Hence, incorporation from injury to injury would be expected to vary.

There are a number of technical reasons why the number of BMD cells might be over- or underestimated. Most studies use a genetic tracer requiring protein expression to follow cell fate. Transgene expression is rarely uniform no matter what the promoter, so underestimation of cell numbers can occur due to loss of gene expression. Even if the transgene is relatively uniformly expressed in hematopoietic cells, it is not valid to assume that the same will be true of their EC progeny. Many studies use GFP to follow the cells. This is extremely problematic. GFP is toxic to cells and can alter their program of gene expression.[164,165] It is also difficult to detect, particularly if one wishes to co-localize it with another antigenic marker. These problems can lead both to over- and underestimation of $GFP^+$ cells numbers. LacZ is not without problems either. It too can be difficult to detect and endogenous enzyme activities can lead to false positives. Moreover different methods of detection can lead to widely different results, but which method more accurately indicates the enzyme's presence is debatable.[150,151] Even techniques that do not rely on protein expression are not infallible. *In situ* hybridization to detect Y chromosomes in sex mismatched transplants is useful, but the false negative rate is 40-60%.[166,167] Still this should yield errors of no more than about two-fold. However, in our experience the time of fixation, age of tissue block, and freshness of slide preparation can all affect sensitivity in *in situ* hybridization. Another *in situ* technique that detects genomic pBR322 sequences that have been integrated into the DNA of transgenic mice is highly sensitive, with false negative rates of less than 9%.[142] However, the procedure is technically difficult, extremely time consuming, and tissue preparation requirements limit the antibodies that can be used to identify specific cell types.

Despite these many limitations, investigators, of course, continue to examine BMD cell contributions to the endothelium. In the first of such studies, BMD cells were found in the endothelium of injury induced neovessels and of normal physiologic ovarian and uterine neovessels, though a quantitative assessment was not performed.[153] Shortly thereafter, as noted above, we reported bone marrow cell contributions to the neovasculature of ~10% in a sponge implant and 1-2% in vessels of uninjured tissues.[142] An important caveat is that the donor cells were fetal liver cells, which may have influenced endothelial engraftment. A subsequent transplant study demonstrated that 3.3% of all vessels in the infarcted heart are bone marrow-derived, but that the cells were localized principally adjacent to the infarct.[107] the percentage in the peri-infarct region was not reported. In another study, however, 20-25% of capillaries in the central infarct zone were reported to be bone marrow-derived in nude rats injected intravenously with human $CD34^+$ cells after myocardial infarct.[143] Thus, the extent of circulating EC precursor participation in vascular growth and maintenance depends on physiological context.

Although the majority of the data in mice and rats suggest significant contributions of circulating EC precursors to the endothelium, a few studies find the opposite. For example, in two murine bone marrow transplant protocols, one where no injury was introduced and one where it was, BMD cells were not detected in the pre-existing or neo-endothelium.[46,168] In addition, few BMD EC were present in matrigel plugs in NOD/scid mice whose bone marrow had been reconstituted by human bone marrow.[169]

Female to male heart transplants make it possible to examine human chimerism in the heart by Y-chromosome labeling. In one such study 20% of arterioles and 15% of capillaries were principally male-derived in the infarcted heart.[102] In contrast, others reported that less than 0.1% of ECs in the heart are BMD cells.[166,167] Although the fraction increased after infarct, the highest percentage of BMD endothelial cells in the infarcted region among five patients was found to be only 1.6%.[167] A mean frequency of 2% donor-derived ECs were detected in the endothelium of skin and gut of transplant recipients.[170] As with animals, we see wide variability in the numbers, but detection of BMD EC is a consistent feature in vascular beds of humans.

## 7. THE FUSS OVER FUSION

A huge controversy arose several years ago when two reports were published showing that adult cells could fuse with stem cells.[171,172] It was immediately suggested that fusion rather than plasticity underlay the putative transdifferentiation of adult BMD cells. Of course, the fusigenic events described in the published works were observed between adult cells and embryonic stem cells, a notoriously fusigenic cell type. Moreover, the fusion events were extremely rare, and could not possibly account for widespread phenotypic changes seen in adult BM and peripheral blood mononuclear cell cultures. Unfortunately, these reports set the adult stem cell field back considerably as many began to question the existence of BMD and other types of adult stem cells. However, they also forced adult stem cell biologists to consider the possibility of fusion, and this has led to some surprising and significant findings. We have learned that adult stem cells both fuse and differentiate. They seem to be most fusigenic in tissues in which fusion has long been known to occur, such as the liver and skeletal muscle.[161,173-175] Fusion also tends to be prevalent in animal models where the endogenous cells are defective, that is, where a functional benefit might be derived from fusion.[21,173-175] On the other hand, in situations where BMD cells are recruited to repair acute damage in otherwise healthy tissue, fusion rarely occurs.[175,176] To our knowledge, there has been only one report of fusion for EC precursors, and in that report only 8% of BMD EC appeared to be derived from fusion.[177] In another report none of >4,000 BMD ECs appeared to have arisen from fusion.[170]

## 8. RE-ENDOTHELIALIZATION, TRANSPLANT ATHEROSCLEROSIS, AND RESTENOSIS

A potential benefit of BMD cells in the vasculature is their ability to promote re-endothelialization. Rapid re-endothelialization can reduce neo-intimal formation and has long been a therapeutic goal. Recent studies show that local introduction of BMD cells can promote re-endothelialization and inhibit neointima formation.[178,179] Interestingly, monocyte chemoattractant protein-1 promoted this process.[178] In addition statins, which appear to mobilize EC precursors,[126,127,151] promote re-endothelialization of balloon injured arteries by recruiting BMD ECs to the site of injury.[180,181] On the other hand, BMD cells are not always helpful, as transplantation of bone marrow from wild type mice into Apo E-/- mice leads to accelerated atherosclerosis.[182]

BMD ECs may also influence transplant atherosclerosis. Among human kidney transplant patients, Y-chromosome labeling or HLA-typing revealed extensive replacement of the graft endothelium by recipient-derived endothelium, with the highest levels of recipient-derived ECs associated with the greatest vascular rejection.[183] Recipient cells also colonized (1-24%) vessels in the hearts of cardiac allograft patients with vasculopathy.[184] In a mouse model, the endothelium of aortic allografts was replaced by host ECs of which 35% were BMD ECs.[185] Interestingly, the BMD EC also made a significant contribution to the microvessels within the lesions. Similarly, when aortic allografts were performed in rats, the endothelium of the grafts was replaced by recipient ECs. However, if transplantation was accompanied by immuno-suppression, the EC were donor-derived.[186] (Of course, it is not known if the recipient EC invasion is the cause or the result of inflammation and rejection, since it is not known whether infiltration precedes or follows endothelial chimerism.) An elegant follow-up study used transplants of aortas from Dark Agouti rats into Brown Norway rats whose bone marrow had been reconstituted with Lewis rat bone marrow.[187] The mice developed significant transplant atherosclerosis, but the endothelium was replaced by host, not BMD cells. In the reciprocal experiment of transplanting Brown Norway bone marrow into Lewis rats, little transplant atherosclerosis developed, but both donor-derived (Dark Agouti) and host-derived ECs were found lining the aorta as well as small number (1-3%) of BMD ECs. This suggests a non-bone marrow source for ECs in the lesions. The introduction of three different sets of histocompatibility antigens into a single animal may have affected the results, but it at least suggests that BMD may not be important in this process in some instances. The significance of host BMD ECs may be organ (or endothelium type) and injury type dependent.

Though clinical trials are underway using BMD cells, a number of reports suggest a detrimental effect of BMD cell therapy on clinical outcomes. Intracoronary injection of $CD133^+$ cells has been associated with accelerated coronary artery atherosclerosis.[188] Mobilization with G-CSF also worsened acute renal failure, albeit probably not through EC precursor mediated effects, but rather due to the concomitant mobilization of granulocytes.[189] Nevertheless, this point is significant, since clinical trials have begun to use mobilizing agents to promote vascularization after myocardial infarction. Recently, G-CSF therapy in conjunction with intra-coronary infusion of autologous mobilized peripheral blood mononuclear cells was found to improve cardiac function, but the trial was stopped because due to an unexpectedly high rate of in-stent restenosis[190]. Clearly, mobilization might not be an appropriate strategy in transplant, CABG, or stented patients.

## 9. BONE MARROW-DERIVED CELLS AS THERAPEUTIC CATALYSTS

The recognition that circulating EC precursors integrate into the endothelium immediately drew attention to the possibility that they might be harnessed to promote vessel growth in ischemic tissues. In 1999 Tomita and colleagues injected freshly isolated or cultured whole bone marrow mononuclear cells into cryo-injured myocardium. Both fresh and cultured cells more than doubled capillary density in the scar relative to controls and improved myocardial function.[191] In the following year a handful of reports also described the benefits of BMD cells in a variety of rodent models in ischemic tissue.[98,109,122,130,192,193] Among the early reports were studies documenting functional improvements in left ventricular developed pressure, ejection fraction, and muscle perfusion after injecting BMD cells into the infarcted rat heart.[193-195] In ischemic hindlimbs, BMD cell injection elevated muscle blood flow and rat limb functional capacity in rats,[196] and dramatically increased limb blood flow and decreased limb loss in mice.[130] Angiographic score and capillary density also improved in ischemic hindlimbs of rabbits injected with BMD cells.[197] Since 2000, work from well over one hundred studies have been published documenting the therapeutic benefits of various subsets of BMD cells either after bone marrow transplantation or systemic or local injection of BMD cells in humans, dogs, cats, rabbits, rats, mice, pigs, and primates.[194,196,198-203]

Hindlimb ischemia is typically induced by femoral artery ligation. In most mouse strains, this results in severe ischemia, necrosis, and often auto-amputation. In some strains of athymic mice, however, femoral artery ligation results in less severe ischemia with only rare toe necrosis and no auto-amputation. Working with these nude mice, we found that local injection of $CD34^+$ peripheral blood mononuclear cells had no effect on blood flow restoration to the ischemic limbs.[109] In contrast, in athymic mice with chemically induced diabetes, the same cells dramatically accelerated the restoration of blood flow to the ischemic limb. Thus, both the systemic and local environments seem to influence the responsiveness of a tissue to BMD cell therapy.

It is clear that the ability of exogenous BMD cells to improve blood flow, enhance capillary growth, and restore tissue function is not restricted to a single cell type. Many mutually exclusive subsets of cells induce vascularization of ischemic tissue. In fact, injection of platelets alone was reported to be sufficient to promote vascularization of an ischemic rat limb.[204] Of course, the effects of the cells are not always equivalent,[133,178] and little is known about which cells under which particular conditions will be most therapeutically efficacious.

Collectively, therapeutic studies show that when exogenous cells are injected locally, the cells may integrate into the endothelium of the ischemic tissue, but do not do so in uninjured tissues. This is good from a safety standpoint. However, in all but a minority of studies, locally injected cells are also rarely found in the endothelium of the ischemic tissue, and typically incorporate into the neo-endothelium at much lower rates than do endogenous BMD cells. This suggests that in order to integrate into the vasculature, a priming step is required, and that cells in the bone marrow are more sensitive to the priming stimulus than those in the tissue or circulation. In support of this, whereas MCP-1 stimulated adhesion and differentiation of bone marrow-derived CD14$^+$ cells, its effects on peripheral blood CD14$^+$ cells was weak.[178] Further, bone marrow, but not blood-derived, CD14$^+$ cells re-endothelialized balloon injured carotid arteries *in vivo*.[178] Also apropos to this, acute myocardial appears to mobilize hemangioblastic EC precursors into the circulation.[145,205-207] Concurrently, mRNA levels of the cardiac muscle antigens GATA4, MEF2c, and Nkx2.5 and of the EC antigens CD144 and vWF are elevated in peripheral blood mononuclear cells.[206] Perhaps, the injured tissue signals the bone marrow to upregulate cell type specific markers, before release into the circulation. Alternatively, tissue specific precursor cells may normally reside in the bone marrow, and they are recruited by factors released by their corresponding tissue during times of need.

Because integration of injected exogenous BMD is rare, the injected cells cannot account for the observed increases in blood flow and capillary density; that is, the functional improvements have little to do with the ability of BMD to differentiate into ECs. How then does one account for the therapeutic effects? Obviously, the most likely scenario is that the injected cells secrete factors that promote healing. Certainly the cells secrete many pro-angiogenic factors that could act directly on endothelial cells, including VEGF.[208,209] Yet, small numbers of locally injected cells influence relatively large masses of tissue. For example, we routinely inject 2.5 x$10^5$ Sca-1$^+$ cells adjacent to 6mm skin wounds in mice, and 5 x $10^5$ human CD34$^+$ cells into ischemic limbs of athymic mice. These few cells increase vascularization throughout the wound, and promote blood flow to the entire ischemic limb, despite the fact that the cells do not appear to migrate more than a short distance from the injection site. This suggests that BMD cells may act as catalysts, stimulating a few local resident, or possibly blood cells, to initiate a positive feedback cascade.

What the molecular mediators of such a cascade might be are unclear, but one molecule in particular, stromal cell-derived factor 1 (SDF-1) bears mention, as it is likely to be a key player. SDF-1 mobilizes hematopoietic stem and progenitor cells from the bone marrow,[210,211] and SDF-1 is upregulated in tissue hypoxia via HIF-1, such that SDF-1 levels follow the ischemic gradient.[212] The SDF-1, which is chemotactic for hemangioblastic EC precursors and EPCs,[132,213,214] in turn is required for recruitment and adhesion of BMD EC precursors to sites of injury.[132,212] Further, it induces differentiation of hemangioblastic EC precursors,[132] and can potentiate tube formation *in vitro*.[215] Additional evidence for a role of SDF-1 in modulating EC precursor function comes from studies on diabetic retinopathy in mice, where SDF-1 is both necessary and sufficient to promote proliferative retinopathy.[216] Blockade of SDF-1 can inhibit retinal vascular growth even in the presence of high levels of VEGF. It is noteworthy that in this model of retinopathy, BMD cells account for a significant proportion of cells in the retinal vasculature.

## 10. DISEASE AND THE FUTURE OF ENDOTHELIAL CELL PRECURSORS

It is becoming clear that there is a relationship between physiologic state and circulating EC precursors. Chronic disease decreases the number of circulating EC precursors, depresses their ability to produce ECs *in vitro*, and reduces their ability to potentiate neovascularization.[109,126,128,139,217-220] Diabetes provides a good example of disease induced EC precursor dysfunction. Human $CD34^+$ cells induce vascular growth in diabetic, but not non-diabetic mice in our athymic mouse model, suggesting BMD EC precursor dysfunction in the diabetic animals.[109] Consistent with this, the EC producing capability of Type 1 diabetic-derived $CD34^+$ cells is reduced.[109] In addition, the normal hypoxia induced increase in EC numbers is ablated in cultures of diabetic-derived mouse $Sca\text{-}1^+$ cells, and diabetic cells exhibit increased sensitivity to oxidative stress induced cell death.[139,217] Further, not only do the diabetic-derived cells have a reduced ability to promote vascular growth, in some instances actually inhibit it.[128,139,217]

While these studies indicate that patients with a variety of chronic diseases are prime candidates for BMD EC therapy, they also suggest that they may be the least likely to benefit from it. Current therapies are designed to use autologous BMD cells. However, intrinsic dysfunction may render these cells marginally efficacious, while exposing the patient to a number of risks, including inhibition of vascular growth by their own cells. Still, ultimately, as we learn more about the regulation of BMD EC precursors and better understand their normal physiological functions, it is probable that strategies will be devised to circumvent endogenous cell dysfunction. BMD EC precursors either by themselves, in combination with pro-angiogenic factors, or as gene therapy vectors will likely become important therapeutic tools in the not too distant future.

## ACKNOWLEDGEMENTS

The author would like to thank Trent Bender for assistance in managing references. This work was supported by grants from the National Institutes of Health (DK 55965 and DK52293 to GCS).

## REFERENCES

1. Till JE: A direct measurement of the radiation sensitivity of normal mouse bone marrow cells. Radiat Res 1961, 14:213-219
2. Mauro A: Satellite cell of skeletal muscle fibers. J Biophys Biochem Cytol 1961, 9:493-495
3. Lemischka IR, Raulet DH, Mulligan RC: Developmental potential and dynamic behavior of hematopoietic stem cells. Cell 1986, 45:917-927
4. Uchida N, Fleming WH, Alpern EJ, Weissman IL: Heterogeneity of hematopoietic stem cells. Curr Opin Immunol 1993, 5:177-184
5. Bischoff R: Regeneration of single skeletal muscle fibers in vitro. Anat Rec 1975, 182:215-235
6. Friedenstein AJ, Gorskaja JF, Kulagina NN: Fibroblast precursors in normal and irradiated mouse hematopoietic organs. Experimental Hematology 1976, 4:267-274
7. Grigoriadis AE, Heersche JN, Aubin JE: Differentiation of muscle, fat, cartilage, and bone from progenitor cells present in a bone-derived clonal cell population: effect of dexamethasone. J Cell Biol 1988, 106:2139-2151
8. Umezawa A, Maruyama T, Segawa K, Shadduck RK, Waheed A, Hata J: Multipotent marrow stromal cell line is able to induce hematopoiesis in vivo. J Cell Physiol 1992, 151:197-205
9. Pereira RF, Halford KW, O'Hara MD, Leeper DB, Sokolov BP, Pollard MD, Bagasra O, Prockop DJ: Cultured adherent cells from marrow can serve as long-lasting precursor cells for bone, cartilage, and lung in irradiated mice. Proc Natl Acad Sci U S A 1995, 92:4857-4861
10. Pittenger MF, Mackay AM, Beck SC, Jaiswal RK, Douglas R, Mosca JD, Moorman MA, Simonetti DW, Craig S, Marshak DR: Multilineage potential of adult human mesenchymal stem cells. Science 1999, 284:143-147
11. Potten CS, Hendry JH: Differential regeneration of intestinal proliferative cells and cryptogenic cells after irradiation. Int J Radiat Biol Relat Stud Phys Chem Med 1975, 27:413-424
12. Potten CS, Loeffler M: Stem cells: attributes, cycles, spirals, pitfalls and uncertainties. Lessons for and from the crypt. Development 1990, 110:1001-1020
13. Kligman AM: The human hair cycle. J Invest Dermatol 1959, 33:307-316
14. Potten CS: The epidermal proliferative unit: the possible role of the central basal cell. Cell Tissue Kinet 1974, 7:77-88
15. Cotsarelis G, Sun TT, Lavker RM: Label-retaining cells reside in the bulge area of pilosebaceous unit: implications for follicular stem cells, hair cycle, and skin carcinogenesis. Cell 1990, 61:1329-1337
16. Asahara T, Murohara T, Sullivan A, Silver M, van der Zee R, Li T, Witzenbichler B, Schatteman G, Isner JM: Isolation of putative progenitor endothelial cells for angiogenesis. Science 1997, 275:964-967
17. Bjornson CR, Rietze RL, Reynolds BA, Magli MC, Vescovi AL: Turning brain into blood: a hematopoietic fate adopted by adult neural stem cells in vivo. Science 1999, 283:534-537
18. Johansson CB, Momma S, Clarke DL, Risling M, Lendahl U, Frisen J: Identification of a neural stem cell in the adult mammalian central nervous system. Cell 1999, 96:25-34
19. Kuhn HG, Svendsen CN: Origins, functions, and potential of adult neural stem cells. Bioessays 1999, 21:625-630

20. Petersen BE, Bowen WC, Patrene KD, Mars WM, Sullivan AK, Murase N, Boggs SS, Greenberger JS, Goff JP: Bone marrow as a potential source of hepatic oval cells. Science 1999, 284:1168-1170
21. Ferrari G, Cusella-De Angelis G, Coletta M, Paolucci E, Stornaiuolo A, Cossu G, Mavilio F: Muscle regeneration by bone marrow-derived myogenic progenitors [published erratum appears in Science 1998 Aug 14;281(5379):923]. Science 1998, 279:1528-1530
22. Bittner RE, Schofer C, Weipoltshammer K, Ivanova S, Streubel B, Hauser E, Freilinger M, Hoger H, Elbe-Burger A, Wachtler F: Recruitment of bone-marrow-derived cells by skeletal and cardiac muscle in adult dystrophic mdx mice. Anat Embryol (Berl) 1999, 199:391-396
23. Makino S, Fukuda K, Miyoshi S, Konishi F, Kodama H, Pan J, Sano M, Takahashi T, Hori S, Abe H, Hata J, Umezawa A, Ogawa S: Cardiomyocytes can be generated from marrow stromal cells in vitro. J Clin Invest 1999, 103:697-705
24. McKinney-Freeman SL, Jackson KA, Camargo FD, Ferrari G, Mavilio F, Goodell MA: Muscle-derived hematopoietic stem cells are hematopoietic in origin. Proc Natl Acad Sci U S A 2002, 99:1341-1346
25. Krause DS, Theise ND, Collector MI, Henegariu O, Hwang S, Gardner R, Neutzel S, Sharkis SJ: Multi-organ, multi-lineage engraftment by a single bone marrow-derived stem cell. Cell 2001, 105:369-377
26. Geiger H, Van Zant G: The aging of lympho-hematopoietic stem cells. Nat Immunol 2002, 3:329-333
27. Globerson A: Haematopoietic stem cell ageing. Novartis Found Symp 2001, 235:85-96
28. Geiger H, True JM, de Haan G, Van Zant G: Age- and stage-specific regulation patterns in the hematopoietic stem cell hierarchy. Blood 2001, 98:2966-2972
29. Suda T, Suda J, Ogawa M: Proliferative kinetics and differentiation of murine blast cell colonies in culture: evidence for variable G0 periods and constant doubling rates of early pluripotent hemopoietic progenitors. J Cell Physiol 1983, 117:308-318
30. Dormer P, Ucci G: Pluripotent stem cells do not completely maintain normal human steady-state haemopoiesis. Cell Tissue Kinet 1984, 17:367-374
31. Loeffler M, Herkenrath P, Wichmann HE, Lord BI, Murphy MJ, Jr.: The kinetics of hematopoietic stem cells during and after hypoxia. A model analysis. Blut 1984, 49:427-439
32. Bickenbach JR, Mackenzie IC: Identification and localization of label-retaining cells in hamster epithelia. J Invest Dermatol 1984, 82:618-622
33. Mackenzie IC, Bickenbach JR: Label-retaining keratinocytes and Langerhans cells in mouse epithelia. Cell Tissue Res 1985, 242:551-556
34. Cheng T, Rodrigues N, Shen H, Yang Y, Dombkowski D, Sykes M, Scadden DT: Hematopoietic stem cell quiescence maintained by p21cip1/waf1. Science 2000, 287:1804-1808
35. Weissman IL: Stem cells: units of development, units of regeneration, and units in evolution. Cell 2000, 100:157-168
36. Dick JE, Bhatia M, Gan O, Kapp U, Wang JC: Assay of human stem cells by repopulation of NOD/SCID mice. Stem Cells 1997, 15 Suppl 1:199-203; discussion 204-197
37. Iscove NN, Nawa K: Hematopoietic stem cells expand during serial transplantation in vivo without apparent exhaustion. Curr Biol 1997, 7:805-808
38. Migliaccio AR, Carta C, Migliaccio G: In vivo expansion of purified hematopoietic stem cells transplanted in nonablated W/Wv mice. Exp Hematol 1999, 27:1655-1666

39. Eichmann A, Corbel C, Nataf V, Vaigot P, Breant C, Le Douarin NM: Ligand-dependent development of the endothelial and hemopoietic lineages from embryonic mesodermal cells expressing vascular endothelial growth factor receptor 2. Proc Natl Acad Sci U S A 1997, 94:5141-5146
40. Choi K, Kennedy M, Kazarov A, Papadimitriou JC, Keller G: A common precursor for hematopoietic and endothelial cells. Development 1998, 125:725-732
41. Tomanek RJ, Schatteman GC: Angiogenesis: new insights and therapeutic potential. Anat Rec 2000, 261:126-135
42. Kurz H, Christ B: Embryonic CNS macrophages and microglia do not stem from circulating, but from extravascular precursors. Glia 1998, 22:98-102
43. Risau W, Flamme I: Vasculogenesis. Annu Rev Cell Dev Biol 1995, 11:73-91
44. Grant MB, May WS, Caballero S, Brown GA, Guthrie SM, Mames RN, Byrne BJ, Vaught T, Spoerri PE, Peck AB, Scott EW: Adult hematopoietic stem cells provide functional hemangioblast activity during retinal neovascularization. Nat Med 2002, 8:607-612
45. Bailey AS, Jiang S, Afentoulis M, Baumann CI, Schroeder DA, Olson SB, Wong MH, Fleming WH: Transplanted adult hematopoietic stems cells differentiate into functional endothelial cells. Blood 2004, 103:13-19
46. Wagers AJ, Sherwood RI, Christensen JL, Weissman IL: Little evidence for developmental plasticity of adult hematopoietic stem cells. Science 2002, 297:2256-2259
47. Scott SM, Barth MG, Gaddy LR, Ahl ET, Jr.: The role of circulating cells in the healing of vascular prostheses. J Vasc Surg 1994, 19:585-593
48. Hueper WC, Russell MA: Capillary-like formations in tissue culture of leukocytes. Arch Exp Zellforsch 1932, 12:407-424
49. Parker RC: The development of organized vessels in cultures of blood cells. Science 1933, 77:544-546
50. White GC, Parshley MS: Growth of in vitro blood vessels from bone marrow of adult chickens. American Journal of Anatomy 1950, 89:321-345
51. Feigl W, Susani M, Ulrich W, Matejka M, Losert U, Sinzinger H: Organisation of experimental thrombosis by blood cells. Virchows Archiv - A 1985, 406:133-148
52. Leu HJ, Feigl W, Susani M: Angiogenesis from mononuclear cells in thrombi. Virchows Arch A Pathol Anat Histopathol 1987, 411:5-14
53. Stump MM, Jordan JL, DeBakey ME, Halpert B: Endothelium grown from circulating blood on isolated intravascular Dacron hub. Am J Pathol 1963, 43:361-367
54. Mackenzie JR, Hackett M, Topuzlu C, Tibbs DJ: Origin of arterial prosthesis lining from circulating blood cells. Arch Surg 1968, 97:879-885
55. Shi Q, Wu MH, Hayashida N, Wechezak AR, Clowes AW, Sauvage LR: Proof of fallout endothelialization of impervious Dacron grafts in the aorta and inferior vena cava of the dog. J Vasc Surg 1994, 20:546-556; discussion 556-547
56. Wu MH, Shi Q, Wechezak AR, Clowes AW, Gordon IL, Sauvage LR: Definitive proof of endothelialization of a Dacron arterial prosthesis in a human being. J Vasc Surg 1995, 21:862-867
57. Noishiki Y, Tomizawa Y, Yamane Y, Matsumoto A: Autocrine angiogenic vascular prosthesis with bone marrow transplantation. Nat Med 1996, 2:90-93
58. Gerrity RG, Caplan BA, Richardson M, Cade JF, Hirsh J, Schwartz CJ: Endotoxin-induced endothelial injury and repair. I. Endothelial cell turnover in the aorta of the rabbit. Exp Mol Pathol 1975, 23:379-385
59. Hladovec J: Circulating endothelial cells as a sign of vessel wall lesions. Physiol Bohemoslov 1978, 27:140-144

60. Sbarbati R, de Boer M, Marzilli M, Scarlattini M, Rossi G, van Mourik JA: Immunologic detection of endothelial cells in human whole blood. Blood 1991, 77:764-769
61. Janssens D, Michiels C, Guillaume G, Cuisinier B, Louagie Y, Remacle J: Increase in circulating endothelial cells in patients with primary chronic venous insufficiency: protective effect of Ginkor Fort in a randomized double-blind, placebo-controlled clinical trial. J Cardiovasc Pharmacol 1999, 33:7-11
62. Mancuso P, Burlini A, Pruneri G, Goldhirsch A, Martinelli G, Bertolini F: Resting and activated endothelial cells are increased in the peripheral blood of cancer patients. Blood 2001, 97:3658-3661
63. Del Papa N, Colombo G, Fracchiolla N, Moronetti LM, Ingegnoli F, Maglione W, Comina DP, Vitali C, Fantini F, Cortelezzi A: Circulating endothelial cells as a marker of ongoing vascular disease in systemic sclerosis. Arthritis Rheum 2004, 50:1296-1304
64. Hladovec J, Prerovsky I, Stanek V, Fabian J: Circulating endothelial cells in acute myocardial infarction and angina pectoris. Klin Wochenschr 1978, 56:1033-1036
65. Prerovsky I, Hladovec J: Suppression of the desquamating effect of smoking on the human endothelium by hydroxyethylrutosides. Blood Vessels 1979, 16:239-240
66. George F, Brisson C, Poncelet P, Laurent JC, Massot O, Arnoux D, Ambrosi P, Klein-Soyer C, Cazenave JP, Sampol J: Rapid isolation of human endothelial cells from whole blood using S-Endo1 monoclonal antibody coupled to immuno-magnetic beads: demonstration of endothelial injury after angioplasty. Thromb Haemost 1992, 67:147-153
67. Grefte A, van der Giessen M, van Son W, The TH: Circulating cytomegalovirus (CMV)-infected endothelial cells in patients with an active CMV infection. J Infect Dis 1993, 167:270-277
68. Sinzinger H, Fitscha P, Kritz H, Rogatti W, Grady JO: Prostaglandin E1 decreases circulating endothelial cells. Prostaglandins 1996, 51:61-68
69. Mutin M, Canavy I, Blann A, Bory M, Sampol J, Dignat-George F: Direct evidence of endothelial injury in acute myocardial infarction and unstable angina by demonstration of circulating endothelial cells. Blood 1999, 93:2951-2958
70. Solovey A, Lin Y, Browne P, Choong S, Wayner E, Hebbel RP: Circulating activated endothelial cells in sickle cell anemia. N Engl J Med 1997, 337:1584-1590
71. Solovey A, Gui L, Ramakrishnan S, Steinberg MH, Hebbel RP: Sickle cell anemia as a possible state of enhanced anti-apoptotic tone: survival effect of vascular endothelial growth factor on circulating and unanchored endothelial cells. Blood 1999, 93:3824-3830
72. Sowemimo-Coker SO, Meiselman HJ, Francis RB, Jr.: Increased circulating endothelial cells in sickle cell crisis. Am J Hematol 1989, 31:263-265
73. Rajagopalan S, Somers EC, Brook RD, Kehrer C, Pfenninger D, Lewis E, Chakrabarti A, Richardson BC, Shelden E, McCune WJ, Kaplan MJ: Endothelial cell apoptosis in systemic lupus erythematosus: a common pathway for abnormal vascular function and thrombosis propensity. Blood 2004, 103:3677-3683
74. Blann AD, Woywodt A, Bertolini F, Bull TM, Buyon JP, Clancy RM, Haubitz M, Hebbel RP, Lip GY, Mancuso P, Sampol J, Solovey A, Dignat-George F: Circulating endothelial cells. Biomarker of vascular disease. Thromb Haemost 2005, 93:228-235
75. Mancuso P, Rabascio C, Bertolini F: Strategies to investigate circulating endothelial cells in cancer. Pathophysiol Haemost Thromb 2003, 33:503-506
76. Mancuso P, Calleri A, Cassi C, Gobbi A, Capillo M, Pruneri G, Martinelli G, Bertolini F: Circulating endothelial cells as a novel marker of angiogenesis. Adv Exp Med Biol 2003, 522:83-97

77. Lin Y, Weisdorf DJ, Solovey A, Hebbel RP: Origins of circulating endothelial cells and endothelial outgrowth from blood. J Clin Invest 2000, 105:71-77
78. Bardin N, George F, Mutin M, Brisson C, Horschowski N, Frances V, Lesaule G, Sampol J: S-Endo 1, a pan-endothelial monoclonal antibody recognizing a novel human endothelial antigen. Tissue Antigens 1996, 48:531-539
79. Solovey AN, Gui L, Chang L, Enenstein J, Browne PV, Hebbel RP: Identification and functional assessment of endothelial P1H12. J Lab Clin Med 2001, 138:322-331
80. Yin AH, Miraglia S, Zanjani ED, Almeida-Porada G, Ogawa M, Leary AG, Olweus J, Kearney J, Buck DW: AC133, a novel marker for human hematopoietic stem and progenitor cells. Blood 1997, 90:5002-5012
81. Miraglia S, Godfrey W, Yin AH, Atkins K, Warnke R, Holden JT, Bray RA, Waller EK, Buck DW: A novel five-transmembrane hematopoietic stem cell antigen: isolation, characterization, and molecular cloning. Blood 1997, 90:5013-5021
82. Sunderland CA, McMaster WR, Williams AF: Purification with monoclonal antibody of a predominant leukocyte-common antigen and glycoprotein from rat thymocytes. Eur J Immunol 1979, 9:155-159
83. Thomas ML: The leukocyte common antigen family. Annu Rev Immunol 1989, 7:339-369
84. Civin CI, L.C. Strauss, C. Brovall, M.J. Fackler, J.F. Schwartz,and J.H. Shaper.: Antigenic analysis of hematopoiesis. III. A hematopoietic progenitor cell surface antigen defined by a monoclonal antibody raised against KG-1a cells. J Immunol 1984, 33:157-165
85. Siena S, Bregni M, Brando B, Ravagnani F, Bonadonna G, Gianni AM: Circulation of CD34+ hematopoietic stem cells in the peripheral blood of high-dose cyclophosphamide-treated patients: enhancement by intravenous recombinant human granulocyte-macrophage colony-stimulating factor. Blood 1989, 74:1905-1914
86. Fritsch G, Stimpfl M, Kurz M, Leitner A, Printz D, Buchinger P, Hoecker P, Gadner H: Characterization of hematopoietic stem cells. Ann N Y Acad Sci 1995, 770:42-52
87. Young PE, Baumhueter S, Lasky LA: The sialomucin CD34 is expressed on hematopoietic cells and blood vessels during murine development. Blood 1995, 85:96-105
88. Krause DS, Fackler MJ, Civin CI, May WS: CD34: structure, biology, and clinical utility. Blood 1996, 87:1-13
89. Newman PJ, Berndt MC, Gorski J, White GCd, Lyman S, Paddock C, Muller WA: PECAM-1 (CD31) cloning and relation to adhesion molecules of the immunoglobulin gene superfamily. Science 1990, 247:1219-1222
90. Newman PJ: The biology of PECAM-1. J Clin Invest 1997, 100:S25-29
91. Sato TN, Qin Y, Kozak CA, Audus KL: Tie-1 and tie-2 define another class of putative receptor tyrosine kinase genes expressed in early embryonic vascular system [published erratum appears in Proc Natl Acad Sci U S A 1993 Dec 15;90(24):12056]. Proc Natl Acad Sci U S A 1993, 90:9355-9358
92. Schnurch H, Risau W: Expression of tie-2, a member of a novel family of receptor tyrosine kinases, in the endothelial cell lineage. Development 1993, 119:957-968
93. Stein O, Stein Y: Bovine aortic endothelial cells display macrophage-like properties towards acetylated 125I-labelled low density lipoprotein. Biochim Biophys Acta 1980, 620:631-635
94. Voyta JC, Via DP, Butterfield CE, Zetter BR: Identification and isolation of endothelial cells based on their increased uptake of acetylated-low density lipoprotein. J Cell Biol 1984, 99:2034-2040

95. Jaffe EA, Nachman RL, Becker CG, Minick CR: Culture of human endothelial cells derived from umbilical veins. Identification by morphologic and immunologic criteria. J Clin Invest 1973, 52:2745-2756
96. Lamas S, Marsden PA, Li GK, Tempst P, Michel T: Endothelial nitric oxide synthase: molecular cloning and characterization of a distinct constitutive enzyme isoform. Proc Natl Acad Sci U S A 1992, 89:6348-6352
97. Harraz M, Jiao C, Hanlon HD, Hartley RS, Schatteman GC: Cd34(-) blood-derived human endothelial cell progenitors. Stem Cells 2001, 19:304-312
98. Ikpeazu C, Davidson MK, Halteman D, Browning PJ, Brandt SJ: Donor origin of circulating endothelial progenitors after allogeneic bone marrow transplantation. Biol Blood Marrow Transplant 2000, 6:301-308
99. Shi Q, Rafii S, Wu MH, Wijelath ES, Yu C, Ishida A, Fujita Y, Kothari S, Mohle R, Sauvage LR, Moore MA, Storb RF, Hammond WP: Evidence for circulating bone marrow-derived endothelial cells. Blood 1998, 92:362-367
100. Gunsilius E, Duba HC, Petzer AL, Kahler CM, Grunewald K, Stockhammer G, Gabl C, Dirnhofer S, Clausen J, Gastl G: Evidence from a leukemia model for maintenance of vascular endothelium by bone-marrow-derived endothelial cells. Lancet 2000, 355:1688-1691
101. Gunsilius E: Evidence from a leukemia model for maintenance of vascular endothelium by bone-marrow-derived endothelial cells. Adv Exp Med Biol 2003, 522:17-24
102. Quaini F, Urbanek K, Beltrami AP, Finato N, Beltrami CA, Nadal-Ginard B, Kajstura J, Leri A, Anversa P: Chimerism of the transplanted heart. N Engl J Med 2002, 346:5-15
103. Cogle CR, Wainman DA, Jorgensen ML, Guthrie SM, Mames RN, Scott EW: Adult human hematopoietic cells provide functional hemangioblast activity. Blood 2004, 103:133-135
104. Nakamura Y, Ando K, Chargui J, Kawada H, Sato T, Tsuji T, Hotta T, Kato S: Ex vivo generation of CD34(+) cells from CD34(-) hematopoietic cells. Blood 1999, 94:4053-4059
105. Gallacher L, Murdoch B, Wu DM, Karanu FN, Keeney M, Bhatia M: Isolation and characterization of human CD34(-)Lin(-) and CD34(+)Lin(-) hematopoietic stem cells using cell surface markers AC133 and CD7. Blood 2000, 95:2813-2820
106. Bonnet D: Haematopoietic stem cells. J Pathol 2002, 197:430-440
107. Jackson KA, Majka SM, Wang H, Pocius J, Hartley CJ, Majesky MW, Entman ML, Michael LH, Hirschi KK, Goodell MA: Regeneration of ischemic cardiac muscle and vascular endothelium by adult stem cells. J Clin Invest 2001, 107:1395-1402
108. Peichev M, Naiyer AJ, Pereira D, Zhu Z, Lane WJ, Williams M, Oz MC, Hicklin DJ, Witte L, Moore MA, Rafii S: Expression of VEGFR-2 and AC133 by circulating human CD34(+) cells identifies a population of functional endothelial precursors. Blood 2000, 95:952-958
109. Schatteman GC, Hanlon HD, Jiao C, Dodds SG, Christy BA: Blood-derived angioblasts accelerate blood-flow restoration in diabetic mice. J Clin Invest 2000, 106:571-578
110. Quirici N, Soligo D, Caneva L, Servida F, Bossolasco P, Deliliers GL: Differentiation and expansion of endothelial cells from human bone marrow CD133(+) cells. Br J Haematol 2001, 115:186-194
111. Salven P, Mustjoki S, Alitalo R, Alitalo K, Rafii S: VEGFR-3 and CD133 identify a population of CD34+ lymphatic/vascular endothelial precursor cells. Blood 2003, 101:168-172
112. Nowak G, Karrar A, Holmen C, Nava S, Uzunel M, Hultenby K, Sumitran-Holgersson S: Expression of vascular endothelial growth factor receptor-2 or tie-2 on peripheral

blood cells defines functionally competent cell populations capable of reendothelialization. Circulation 2004, 110:3699-3707
113. Minasi MG, Riminucci M, De Angelis L, Borello U, Berarducci B, Innocenzi A, Caprioli A, Sirabella D, Baiocchi M, De Maria R, Boratto R, Jaffredo T, Broccoli V, Bianco P, Cossu G: The meso-angioblast: a multipotent, self-renewing cell that originates from the dorsal aorta and differentiates into most mesodermal tissues. Development 2002, 129:2773-2783
114. Ema M, Faloon P, Zhang WJ, Hirashima M, Reid T, Stanford WL, Orkin S, Choi K, Rossant J: Combinatorial effects of Flk1 and Tal1 on vascular and hematopoietic development in the mouse. Genes Dev 2003, 17:380-393
115. Robertson SM, Kennedy M, Shannon JM, Keller G: A transitional stage in the commitment of mesoderm to hematopoiesis requiring the transcription factor SCL/tal-1. Development 2000, 127:2447-2459
116. Fernandez Pujol B, Lucibello FC, Gehling UM, Lindemann K, Weidner N, Zuzarte ML, Adamkiewicz J, Elsasser HP, Muller R, Havemann K: Endothelial-like cells derived from human CD14 positive monocytes. Differentiation 2000, 65:287-300
117. Moldovan NI, Goldschmidt-Clermont PJ, Parker-Thornburg J, Shapiro SD, Kolattukudy PE: Contribution of monocytes/macrophages to compensatory neovascularization: the drilling of metalloelastase-positive tunnels in ischemic myocardium. Circ Res 2000, 87:378-384
118. Fernandez Pujol B, Lucibello FC, Zuzarte M, Lutjens P, Muller R, Havemann K: Dendritic cells derived from peripheral monocytes express endothelial markers and in the presence of angiogenic growth factors differentiate into endothelial-like cells. Eur J Cell Biol 2001, 80:99-110
119. Schmeisser A, Garlichs CD, Zhang H, Eskafi S, Graffy C, Ludwig J, Strasser RH, Daniel WG: Monocytes coexpress endothelial and macrophagocytic lineage markers and form cord-like structures in Matrigel under angiogenic conditions. Cardiovasc Res 2001, 49:671-680
120. Rehman J, Li J, Orschell CM, March KL: Peripheral blood "endothelial progenitor cells" are derived from monocyte/macrophages and secrete angiogenic growth factors. Circulation 2003, 107:1164-1169
121. Zhao Y, Glesne D, Huberman E: A human peripheral blood monocyte-derived subset acts as pluripotent stem cells. Proc Natl Acad Sci U S A 2003, 100:2426-2431
122. Murohara T, Ikeda H, Duan J, Shintani S, Sasaki K, Eguchi H, Onitsuka I, Matsui K, Imaizumi T: Transplanted cord blood-derived endothelial precursor cells augment postnatal neovascularization. J Clin Invest 2000, 105:1527-1536
123. Hernandez DA, Townsend LE, Uzieblo MR, Haan ME, Callahan RE, Bendick PJ, Glover JL: Human endothelial cell cultures from progenitor cells obtained by leukapheresis. Am Surg 2000, 66:355-358; discussion 359
124. Asahara T, Takahashi T, Masuda H, Kalka C, Chen D, Iwaguro H, Inai Y, Silver M, Isner JM: VEGF contributes to postnatal neovascularization by mobilizing bone marrow-derived endothelial progenitor cells. Embo J 1999, 18:3964-3972
125. Kalka C, Masuda H, Takahashi T, Gordon R, Tepper O, Gravereaux E, Pieczek A, Iwaguro H, Hayashi SI, Isner JM, Asahara T: Vascular endothelial growth factor(165) gene transfer augments circulating endothelial progenitor cells in human subjects. Circ Res 2000, 86:1198-1202
126. Vasa M, Fichtlscherer S, Adler K, Aicher A, Martin H, Zeiher AM, Dimmeler S: Increase in circulating endothelial progenitor cells by statin therapy in patients with stable coronary artery disease. Circulation 2001, 103:2885-2890

127. Dimmeler S, Aicher A, Vasa M, Mildner-Rihm C, Adler K, Tiemann M, Rutten H, Fichtlscherer S, Martin H, Zeiher AM: HMG-CoA reductase inhibitors (statins) increase endothelial progenitor cells via the PI 3-kinase/Akt pathway. J Clin Invest 2001, 108:391-397
128. Tepper OM, Galiano RD, Capla JM, Kalka C, Gagne PJ, Jacobowitz GR, Levine JP, Gurtner GC: Human endothelial progenitor cells from type II diabetics exhibit impaired proliferation, adhesion, and incorporation into vascular structures. Circulation 2002, 106:2781-2786
129. Loomans CJM, de Koning EJP, Staal FJT, Rookmaaker MB, Verseyden C, de Boer HC, Verhaar MC, Braam B, Rabelink TJ, van Zonneveld AJ: Endothelial progenitor cell dysfunction - A novel concept in the pathogenesis of vascular complications of type 1 diabetes. Diabetes 2004, 53:195-199
130. Kalka C, Masuda H, Takahashi T, Kalka-Moll WM, Silver M, Kearney M, Li T, Isner JM, Asahara T: Transplantation of ex vivo expanded endothelial progenitor cells for therapeutic neovascularization. Proc Natl Acad Sci U S A 2000, 97:3422-3427
131. Nakul-Aquaronne D, Bayle J, Frelin C: Coexpression of endothelial markers and CD14 by cytokine mobilized CD34+ cells under angiogenic stimulation. Cardiovasc Res 2003, 57:816-823
132. De Falco E, Porcelli D, Torella AR, Straino S, Iachininoto MG, Orlandi A, Truffa S, Biglioli P, Napolitano M, Capogrossi MC, Pesce M: SDF-1 involvement in endothelial phenotype and ischemia-induced recruitment of bone marrow progenitor cells. Blood 2004, 104:3472-3482
133. Urbich C, Heeschen C, Aicher A, Dernbach E, Zeiher AM, Dimmeler S: Relevance of monocytic features for neovascularization capacity of circulating endothelial progenitor cells. Circulation 2003, 108:2511-2516.
134. Morel F, Galy A, Chen B, Szilvassy SJ: Equal distribution of competitive long-term repopulating stem cells in the CD34+ and CD34- fractions of Thy-1lowLin-/lowSca-1+ bone marrow cells. Exp Hematol 1998, 26:440-448.
135. Matsuoka S, Ebihara Y, Xu M, Ishii T, Sugiyama D, Yoshino H, Ueda T, Manabe A, Tanaka R, Ikeda Y, Nakahata T, Tsuji K: CD34 expression on long-term repopulating hematopoietic stem cells changes during developmental stages. Blood 2001, 97:419-425.
136. Sogo S, Inaba M, Ogata H, Hisha H, Adachi Y, Mori S, Toki J, Yamanishi K, Kanzaki H, Adachi M, Ikehara S: Induction of c-kit molecules on human CD34+/c-kit < low cells: evidence for CD34+/c-kit < low cells as primitive hematopoietic stem cells. Stem Cells 1997, 15:420-429
137. Xiao M, Oppenlander BK, Plunkett JM, Dooley DC: Expression of Flt3 and c-kit during growth and maturation of human CD34+CD38- cells. Exp Hematol 1999, 27:916-927
138. Sitnicka E, Buza-Vidas N, Larsson S, Nygren JM, Liuba K, Jacobsen SE: Human CD34+ hematopoietic stem cells capable of multilineage engrafting NOD/SCID mice express flt3: distinct flt3 and c-kit expression and response patterns on mouse and candidate human hematopoietic stem cells. Blood 2003, 102:881-886
139. Awad O, Jiao C, Ma N, Dunnwald M, Schatteman GC: Obese diabetic mouse environment differentially affects primitive and monocytic endothelial cell progenitors. Stem Cells 2005, 23: 575-583
140. Wang C, Jiao C, Hanlon HD, Zheng W, Tomanek RJ, Schatteman GC: Mechanical, cellular, and molecular factors interact to modulate circulating endothelial cell progenitors. Am J Physiol Heart Circ Physiol 2004, 286:H1985-1993
141. Gerrity RG, Richardson M, Caplan BA, Cade JF, Hirsh J, Schwartz CJ: Endotoxin-induced vascular endothelial injury and repair. II. Focal injury, en face morphology,

(3H)thymidine uptake and circulating endothelial cells in the dog. Exp Mol Pathol 1976, 24:59-69.

142. Crosby JR, Kaminski WE, Schatteman GC, Martin JC, Raines EW, Seifert RA, Bowen-Pope DF: Endothelial cells of hematopoietic origin make a significant contribution to adult blood vessel formation. Circ Res 2000, 87:728-730
143. Kocher AA, Schuster MD, Szabolcs MJ, Takuma S, Burkhoff D, Wang J, Homma S, Edwards NM, Itescu S: Neovascularization of ischemic myocardium by human bone-marrow-derived angioblasts prevents cardiomyocyte apoptosis, reduces remodeling and improves cardiac function. Nat Med 2001, 7:430-436
144. Takahashi T, Kalka C, Masuda H, Chen D, Silver M, Kearney M, Magner M, Isner JM, Asahara T: Ischemia- and cytokine-induced mobilization of bone marrow-derived endothelial progenitor cells for neovascularization. Nat Med 1999, 5:434-438
145. Shintani S, Murohara T, Ikeda H, Ueno T, Honma T, Katoh A, Sasaki K, Shimada T, Oike Y, Imaizumi T: Mobilization of endothelial progenitor cells in patients with acute myocardial infarction. Circulation 2001, 103:2776-2779
146. Gill M, Dias S, Hattori K, Rivera ML, Hicklin D, Witte L, Girardi L, Yurt R, Himel H, Rafii S: Vascular trauma induces rapid but transient mobilization of VEGFR2(+)AC133(+) endothelial precursor cells. Circ Res 2001, 88:167-174
147. Mathews V, Hanson PT, Ford E, Fujita J, Polonsky KS, Graubert TA: Recruitment of bone marrow-derived endothelial cells to sites of pancreatic beta-cell injury. Diabetes 2004, 53:91-98
148. Laufs U, Werner N, Link A, Endres M, Wassmann S, Jurgens K, Miche E, Bohm M, Nickenig G: Physical training increases endothelial progenitor cells, inhibits neointima formation, and enhances angiogenesis. Circulation 2004, 109:220-226
149. Tepper OM, Capla JM, Galiano RD, Ceradini DJ, Callaghan MJ, Kleinman ME, Gurtner GC: Adult vasculogenesis occurs through in situ recruitment, proliferation, and tubulization of circulating bone marrow-derived cells. Blood 2005, 105:1068-1077
150. Murayama T, Tepper OM, Silver M, Ma H, Losordo DW, Isner JM, Asahara T, Kalka C: Determination of bone marrow-derived endothelial progenitor cell significance in angiogenic growth factor-induced neovascularization in vivo. Exp Hematol 2002, 30:967-972
151. Llevadot J, Murasawa S, Kureishi Y, Uchida S, Masuda H, Kawamoto A, Walsh K, Isner JM, Asahara T: HMG-CoA reductase inhibitor mobilizes bone marrow--derived endothelial progenitor cells. J Clin Invest 2001, 108:399-405
152. Peters BA, Diaz LA, Polyak K, Meszler L, Romans K, Guinan EC, Antin JH, Myerson D, Hamilton SR, Vogelstein B, Kinzler KW, Lengauer C: Contribution of bone marrow-derived endothelial cells to human tumor vasculature. Nat Med 2005, 11:261-262
153. Asahara T, Masuda H, Takahashi T, Kalka C, Pastore C, Silver M, Kearne M, Magner M, Isner JM: Bone marrow origin of endothelial progenitor cells responsible for postnatal vasculogenesis in physiological and pathological neovascularization. Circ Res 1999, 85:221-228
154. Shaw JP, Basch R, Shamamian P: Hematopoietic stem cells and endothelial cell precursors express Tie-2, CD31 and CD45. Blood Cells Mol Dis 2004, 32:168-175
155. Natori T, Sata M, Washida M, Hirata Y, Nagai R, Makuuchi M: G-CSF stimulates angiogenesis and promotes tumor growth: potential contribution of bone marrow-derived endothelial progenitor cells. Biochem Biophys Res Commun 2002, 297:1058-1061
156. Lyden D, Hattori K, Dias S, Costa C, Blaikie P, Butros L, Chadburn A, Heissig B, Marks W, Witte L, Wu Y, Hicklin D, Zhu Z, Hackett NR, Crystal RG, Moore MA, Hajjar KA, Manova K, Benezra R, Rafii S: Impaired recruitment of bone-marrow-

derived endothelial and hematopoietic precursor cells blocks tumor angiogenesis and growth. Nat Med 2001, 7:1194-1201

157. Davidoff AM, Ng CY, Brown P, Leary MA, Spurbeck WW, Zhou J, Horwitz E, Vanin EF, Nienhuis AW: Bone marrow-derived cells contribute to tumor neovasculature and, when modified to express an angiogenesis inhibitor, can restrict tumor growth in mice. Clin Cancer Res 2001, 7:2870-2879
158. Ferrari N, Glod J, Lee J, Kobiler D, Fine HA: Bone marrow-derived, endothelial progenitor-like cells as angiogenesis-selective gene-targeting vectors. Gene Ther 2003, 10:647-656
159. Uchida N, He D, Friera AM, Reitsma M, Sasaki D, Chen B, Tsukamoto A: The unexpected G0/G1 cell cycle status of mobilized hematopoietic stem cells from peripheral blood. Blood 1997, 89:465-472
160. Quesenberry PJ, Becker P, Stewart FM: Phenotype of the engrafting stem cell in mice. Stem Cells 1998, 16 Suppl 1:33-35
161. Camargo FD, Green R, Capetenaki Y, Jackson KA, Goodell MA: Single hematopoietic stem cells generate skeletal muscle through myeloid intermediates. Nat Med 2003, 9:1520-1527
162. Doyonnas R, LaBarge MA, Sacco A, Charlton C, Blau HM: Hematopoietic contribution to skeletal muscle regeneration by myelomonocytic precursors. Proc Natl Acad Sci U S A 2004, 101:13507-13512
163. Brazelton TR, Nystrom M, Blau HM: Significant differences among skeletal muscles in the incorporation of bone marrow-derived cells. Developmental Biology 2003, 262:64-74
164. Liu HS, Jan MS, Chou CK, Chen PH, Ke NJ: Is green fluorescent protein toxic to the living cells? Biochem Biophys Res Commun 1999, 260:712-717
165. Zhang F, Hackett NR, Lam G, Cheng J, Pergolizzi R, Luo L, Shmelkov SV, Edelberg J, Crystal RG, Rafii S: Green fluorescent protein selectively induces HSP70-mediated up-regulation of COX-2 expression in endothelial cells. Blood 2003, 102:2115-2121
166. Murry CE, Soonpaa MH, Reinecke H, Nakajima H, Nakajima HO, Rubart M, Pasumarthi KB, Virag JI, Bartelmez SH, Poppa V, Bradford G, Dowell JD, Williams DA, Field LJ: Haematopoietic stem cells do not transdifferentiate into cardiac myocytes in myocardial infarcts. Nature 2004, 428:664-668
167. Hocht-Zeisberg E, Kahnert H, Guan K, Wulf G, Hemmerlein B, Schlott T, Tenderich G, Korfer R, Raute-Kreinsen U, Hassenfuss G: Cellular repopulation of myocardial infarction in patients with sex-mismatched heart transplantation. Eur Heart J 2004, 25:749-758
168. Balsam LB, Wagers AJ, Christensen JL, Kofidis T, Weissman IL, Robbins RC: Hematopoietic stem cells adopt mature hematopoietic fates in ischaemic myocardium. Nature 2004, 428:668-673
169. Droetto S, Viale A, Primo L, Jordaney N, Bruno S, Pagano M, Piacibello W, Bussolino F, Aglietta M: Vasculogenic potential of long term repopulating cord blood progenitors. Faseb J 2004, 18:1273-1275
170. Jiang S, Walker L, Afentoulis M, Anderson DA, Jauron-Mills L, Corless CL, Fleming WH: Transplanted human bone marrow contributes to vascular endothelium. Proc Natl Acad Sci U S A 2004, 101:16891-16896
171. Ying QL, Nichols J, Evans EP, Smith AG: Changing potency by spontaneous fusion. Nature 2002, 416:545-548

172. Terada N, Hamazaki T, Oka M, Hoki M, Mastalerz DM, Nakano Y, Meyer EM, Morel L, Petersen BE, Scott EW: Bone marrow cells adopt the phenotype of other cells by spontaneous cell fusion. Nature 2002, 416:542-545
173. Vassilopoulos G, Wang PR, Russell DW: Transplanted bone marrow regenerates liver by cell fusion. Nature 2003, 422:901-904
174. Wang X, Willenbring H, Akkari Y, Torimaru Y, Foster M, Al-Dhalimy M, Lagasse E, Finegold M, Olson S, Grompe M: Cell fusion is the principal source of bone-marrow-derived hepatocytes. Nature 2003, 422:897-901
175. Harris RG, Herzog EL, Bruscia EM, Grove JE, Van Arnam JS, Krause DS: Lack of a fusion requirement for development of bone marrow-derived epithelia. Science 2004, 305:90-93
176. LaBarge MA, Blau HM: Biological progression from adult bone marrow to mononucleate muscle stem cell to multinucleate muscle fiber in response to injury. Cell 2002, 111:589-601
177. Zhang S, Wang D, Estrov Z, Raj S, Willerson JT, Yeh ET: Both cell fusion and transdifferentiation account for the transformation of human peripheral blood CD34-positive cells into cardiomyocytes in vivo. Circulation 2004, 110:3803-3807
178. Fujiyama S, Amano K, Uehira K, Yoshida M, Nishiwaki Y, Nozawa Y, Jin D, Takai S, Miyazaki M, Egashira K, Imada T, Iwasaka T, Matsubara H: Bone Marrow Monocyte Lineage Cells Adhere on Injured Endothelium in a Monocyte Chmeattractant Protein-1-Dependent Manner and Accelerate Reendothelialization as Endothelial Progenitor Cells. Circ Res 2003, 2003:1-10
179. Gulati R, Jevremovic D, Witt TA, Kleppe LS, Vile RG, Lerman A, Simari RD: Modulation of the vascular response to injury by autologous blood-derived outgrowth endothelial cells. Am J Physiol Heart Circ Physiol 2004, 287:H512-517
180. Walter DH, Rittig K, Bahlmann FH, Kirchmair R, Silver M, Murayama T, Nishimura H, Losordo DW, Asahara T, Isner JM: Statin therapy accelerates reendothelialization: a novel effect involving mobilization and incorporation of bone marrow-derived endothelial progenitor cells. Circulation 2002, 105:3017-3024
181. Werner N, Priller J, Laufs U, Endres M, Bohm M, Dirnagl U, Nickenig G: Bone marrow-derived progenitor cells modulate vascular reendothelialization and neointimal formation: effect of 3-hydroxy-3-methylglutaryl coenzyme a reductase inhibition. Arterioscler Thromb Vasc Biol 2002, 22:1567-1572
182. Silvestre JS, Gojova A, Brun V, Potteaux S, Esposito B, Duriez M, Clergue M, Le Ricousse-Roussanne S, Barateau V, Merval R, Groux H, Tobelem G, Levy B, Tedgui A, Mallat Z: Transplantation of bone marrow-derived mononuclear cells in ischemic apolipoprotein E-knockout mice accelerates atherosclerosis without altering plaque composition. Circulation 2003, 108:2839-2842
183. Lagaaij EL, Cramer-Knijnenburg GF, van Kemenade FJ, van Es LA, Bruijn JA, van Krieken JH: Endothelial cell chimerism after renal transplantation and vascular rejection. Lancet 2001, 357:33-37
184. Simper D, Wang S, Deb A, Holmes D, McGregor C, Frantz R, Kushwaha SS, Caplice NM: Endothelial progenitor cells are decreased in blood of cardiac allograft patients with vasculopathy and endothelial cells of noncardiac origin are enriched in transplant atherosclerosis. Circulation 2003, 108:143-149
185. Hu Y, Davison F, Zhang Z, Xu Q: Endothelial replacement and angiogenesis in arteriosclerotic lesions of allografts are contributed by circulating progenitor cells. Circulation 2003, 108:3122-3127

186. Hillebrands JL, Klatter FA, van den Hurk BM, Popa ER, Nieuwenhuis P, Rozing J: Origin of neointimal endothelium and alpha-actin-positive smooth muscle cells in transplant arteriosclerosis. J Clin Invest 2001, 107:1411-1422
187. Hillebrands JL, Klatter FA, van Dijk WD, Rozing J: Bone marrow does not contribute substantially to endothelial-cell replacement in transplant arteriosclerosis. Nat Med 2002, 8:194-195
188. Vanderheyden M, Mansour S, Bartunek J: Accelerated atherosclerosis following intracoronary haematopoietic stem cell administration. Heart 2005, 91:448
189. Togel F, Isaac J, Westenfelder C: Hematopoietic stem cell mobilization-associated granulocytosis severely worsens acute renal failure. J Am Soc Nephrol 2004, 15:1261-1267
190. Kang HJ, Kim HS, Zhang SY, Park KW, Cho HJ, Koo BK, Kim YJ, Soo Lee D, Sohn DW, Han KS, Oh BH, Lee MM, Park YB: Effects of intracoronary infusion of peripheral blood stem-cells mobilised with granulocyte-colony stimulating factor on left ventricular systolic function and restenosis after coronary stenting in myocardial infarction: the MAGIC cell randomised clinical trial. Lancet 2004, 363:751-756
191. Tomita S, Li RK, Weisel RD, Mickle DA, Kim EJ, Sakai T, Jia ZQ: Autologous transplantation of bone marrow cells improves damaged heart function. Circulation 1999, 100:II247-256
192. Bhattacharya V, McSweeney PA, Shi Q, Bruno B, Ishida A, Nash R, Storb RF, Sauvage LR, Hammond WP, Wu MH: Enhanced endothelialization and microvessel formation in polyester grafts seeded with CD34(+) bone marrow cells. Blood 2000, 95:581-585
193. Kobayashi T, Hamano K, Li TS, Katoh T, Kobayashi S, Matsuzaki M, Esato K: Enhancement of angiogenesis by the implantation of self bone marrow cells in a rat ischemic heart model. J Surg Res 2000, 89:189-195
194. Kamihata H, Matsubara H, Nishiue T, Fujiyama S, Tsutsumi Y, Ozono R, Masaki H, Mori Y, Iba O, Tateishi E, Kosaki A, Shintani S, Murohara T, Imaizumi T, Iwasaka T: Implantation of bone marrow mononuclear cells into ischemic myocardium enhances collateral perfusion and regional function via side supply of angioblasts, angiogenic ligands, and cytokines. Circulation 2001, 104:1046-1052
195. Kawamoto A, Gwon HC, Iwaguro H, Yamaguchi JI, Uchida S, Masuda H, Silver M, Ma H, Kearney M, Isner JM, Asahara T: Therapeutic potential of ex vivo expanded endothelial progenitor cells for myocardial ischemia. Circulation 2001, 103:634-637
196. Ikenaga S, Hamano K, Nishida M, Kobayashi T, Li TS, Kobayashi S, Matsuzaki M, Zempo N, Esato K: Autologous bone marrow implantation induced angiogenesis and improved deteriorated exercise capacity in a rat ischemic hindlimb model. J Surg Res 2001, 96:277-283
197. Shintani S, Murohara T, Ikeda H, Ueno T, Sasaki K, Duan J, Imaizumi T: Augmentation of postnatal neovascularization with autologous bone marrow transplantation. Circulation 2001, 103:897-903
198. Fuchs S, Baffour R, Zhou YF, Shou M, Pierre A, Tio FO, Weissman NJ, Leon MB, Epstein SE, Kornowski R: Transendocardial delivery of autologous bone marrow enhances collateral perfusion and regional function in pigs with chronic experimental myocardial ischemia. J Am Coll Cardiol 2001, 37:1726-1732
199. Hamano L, Kobayashi, Tanaka, Kobayashi, Esato: The induction of angiogenesis by the implantation of autologous bone marrow cells: a novel and simple therapeutic method. Surgery, 2001, pp 44-54
200. Kawamoto A, Tkebuchava T, Yamaguchi J, Nishimura H, Yoon YS, Milliken C, Uchida S, Masuo O, Iwaguro H, Ma H, Hanley A, Silver M, Kearney M, Losordo DW, Isner JM, Asahara T: Intramyocardial transplantation of autologous endothelial

progenitor cells for therapeutic neovascularization of myocardial ischemia. Circulation 2003, 107:461-468
201. Yoshioka T, Ageyama N, Shibata H, Yasu T, Misawa Y, Takeuchi K, Matsui K, Yamamoto K, Terao K, Shimada K, Ikeda U, Ozawa K, Hanazono Y: Repair of infarcted myocardium mediated by transplanted bone marrow-derived CD34+ stem cells in a nonhuman primate model. Stem Cells 2005, 23:355-364
202. Madeddu P, Emanueli C, Pelosi E, Salis MB, Cerio AM, Bonanno G, Patti M, Stassi G, Condorelli G, Peschle C: Transplantation of low dose CD34+KDR+ cells promotes vascular and muscular regeneration in ischemic limbs. Faseb J 2004, 18:1737-1739
203. Ott I, Keller U, Knoedler M, Gotze KS, Doss K, Fischer P, Urlbauer K, Debus G, von Bubnoff N, Rudelius M, Schomig A, Peschel C, Oostendorp RA: Endothelial-like cells expanded from CD34+ blood cells improve left ventricular function after experimental myocardial infarction. Faseb J 2005, 25:749-758
204. Norol F, Merlet P, Isnard R, Sebillon P, Bonnet N, Cailliot C, Carrion C, Ribeiro M, Charlotte F, Pradeau P, Mayol JF, Peinnequin A, Drouet M, Safsafi K, Vernant JP, Herodin F: Influence of mobilized stem cells on myocardial infarct repair in a nonhuman primate model. Blood 2003, 102:4361-4368
205. Massa M, Rosti V, Ferrario M, Campanelli R, Ramajoli I, Rosso R, De Ferrari GM, Ferlini M, Goffredo L, Bertoletti A, Klersy C, Pecci A, Moratti R, Tavazzi L: Increased circulating hematopoietic and endothelial progenitor cells in the early phase of acute myocardial infarction. Blood 2005, 105:199-206
206. Wojakowski W, Tendera M, Michalowska A, Majka M, Kucia M, Maslankiewicz K, Wyderka R, Ochala A, Ratajczak MZ: Mobilization of CD34/CXCR4+, CD34/CD117+, c-met+ stem cells, and mononuclear cells expressing early cardiac, muscle, and endothelial markers into peripheral blood in patients with acute myocardial infarction. Circulation 2004, 110:3213-3220
207. Orlic D, Kajstura J, Chimenti S, Limana F, Jakoniuk I, Quaini F, Nadal-Ginard B, Bodine DM, Leri A, Anversa P: Mobilized bone marrow cells repair the infarcted heart, improving function and survival. Proc Natl Acad Sci U S A 2001, 98:10344-10349
208. Majka M, Janowska-Wieczorek A, Ratajczak J, Ehrenman K, Pietrzkowski Z, Kowalska MA, Gewirtz AM, Emerson SG, Ratajczak MZ: Numerous growth factors, cytokines, and chemokines are secreted by human CD34(+) cells, myeloblasts, erythroblasts, and megakaryoblasts and regulate normal hematopoiesis in an autocrine/paracrine manner. Blood 2001, 97:3075-3085
209. Bautz F, Rafii S, Kanz L, Mohle R: Expression and secretion of vascular endothelial growth factor-A by cytokine-stimulated hematopoietic progenitor cells. Possible role in the hematopoietic microenvironment. Exp Hematol 2000, 28:700-706
210. Hattori K, Heissig B, Tashiro K, Honjo T, Tateno M, Shieh JH, Hackett NR, Quitoriano MS, Crystal RG, Rafii S, Moore MA: Plasma elevation of stromal cell-derived factor-1 induces mobilization of mature and immature hematopoietic progenitor and stem cells. Blood 2001, 97:3354-3360
211. Wright DE, Bowman EP, Wagers AJ, Butcher EC, Weissman IL: Hematopoietic stem cells are uniquely selective in their migratory response to chemokines. J Exp Med 2002, 195:1145-1154
212. Ceradini DJ, Kulkarni AR, Callaghan MJ, Tepper OM, Bastidas N, Kleinman ME, Capla JM, Galiano RD, Levine JP, Gurtner GC: Progenitor cell trafficking is regulated by hypoxic gradients through HIF-1 induction of SDF-1. Nat Med 2004, 10:858-864
213. Moore MA, Hattori K, Heissig B, Shieh JH, Dias S, Crystal RG, Rafii S: Mobilization of endothelial and hematopoietic stem and progenitor cells by adenovector-mediated

elevation of serum levels of SDF-1, VEGF, and angiopoietin-1. Ann N Y Acad Sci 2001, 938:36-45; discussion 45-37

214. Yamaguchi J, Kusano KF, Masuo O, Kawamoto A, Silver M, Murasawa S, Bosch-Marce M, Masuda H, Losordo DW, Isner JM, Asahara T: Stromal cell-derived factor-1 effects on ex vivo expanded endothelial progenitor cell recruitment for ischemic neovascularization. Circulation 2003, 107:1322-1328
215. Salvucci O, Yao L, Villalba S, Sajewicz A, Pittaluga S, Tosato G: Regulation of endothelial cell branching morphogenesis by endogenous chemokine stromal-derived factor-1. Blood 2002, 99:2703-2711
216. Butler JM, Guthrie SM, Koc M, Afzal A, Caballero S, Brooks HL, Mames RN, Segal MS, Grant MB, Scott EW: SDF-1 is both necessary and sufficient to promote proliferative retinopathy. Journal of Clinical Investigation 2005, 115:86-93
217. Stepanovic V, Awad O, Jiao C, Dunnwald M, Schatteman G: Leprdb diabetic mouse bone marrow cells inhibit skin wound vascularization but promote wound healing. Circ Res 2003, 92:1247-1253
218. Heeschen L, Honold, Assmus, Aicher, Walter, Martin, Zeiher, Dimmeler: Profoundly reduced neovascularization capacity of bone marrow mononuclear cells derived from patients with chronic ischemic heart disease. Circulation, 2004, pp 1615-1622
219. Yamamoto K, Kondo T, Suzuki S, Izawa H, Kobayashi M, Emi N, Komori K, Naoe T, Takamatsu J, Murohara T: Molecular evaluation of endothelial progenitor cells in patients with ischemic limbs: therapeutic effect by stem cell transplantation. Arteriosclerosis, thrombosis, and vascular biology 2004, 24:e192-e196
220. Powell TM, Paul JD, Hill JM, Thompson M, Benjamin M, Rodrigo M, McCoy JP, Read EJ, Khuu HM, Leitman SF, Finkel T, Cannon RO, 3rd: Granulocyte colony-stimulating factor mobilizes functional endothelial progenitor cells in patients with coronary artery disease. Arterioscler Thromb Vasc Biol 2005, 25:296-301

Chapter 4

# REGULATION OF POSTANGIOGENIC VASCULAR REGRESSION

Roberto F. Nicosia[1,2], Wen-Hui Zhu[1], Alfred C. Aplin[1]
*Division of Pathology and Laboratory Medicine, VA Puget Sound Health Care System[1], Department of Pathology, University of Washington[2]*

## 1. INTRODUCTION

Angiogenesis is a complex morphogenetic process characterized by multiple mechanisms of vessel wall assembly and remodeling. During primary angiogenesis (vasculogenesis) blood vessels arise *de novo* from a subpopulation of mesodermal cells, which differentiate into a polygonal network of endothelial tubes lacking a mural cell coating. Further expansion of the primitive embryonal vascular system occurs through sprouting of neovessels from the differentiated endothelium of preexisting vessels.[1] Progenitor endothelial cells participate in this secondary form of angiogenesis by becoming incorporated into neovessels from the adjacent perivascular mesenchyme.[2] Capillary loops form through end-to-end fusion of endothelial sprouts or by intussusceptive microvascular growth, a variant of angiogenesis in which transluminal endothelial pillars subdivide the luminal space of an individual sinusoidal capillary into two separate vessels.[3] Neovessels enlarge by intercalated growth of endothelial cells, which proliferate without sprouting, and by lateral fusion of preformed vascular tubes, like the paired dorsal aortas.[2] As blood starts to flow and tissues differentiate, the primary vascular plexus transforms into an arborizing network of arteries, capillaries, and veins. Blood vessel walls thicken by acquiring a coating of mural cells (pericytes and smooth muscle cells) that originate from the mesodermal and neural crest mesenchyme.[4-7]



*R. Forough (ed.), New Frontiers in Angiogenesis,* 79–95.

After birth, blood vessels maintain their plasticity and continue to expand to meet the oxygen and nutritional requirements of the growing organism. Once an individual has matured and reached adulthood the angiogenic process ceases to be active except during ovarian and endometrial cycles[8,9], pregnancy[9,10], and lactation[11], which are all characterized by hormonally regulated phases of florid angiogenic activity. Endothelial cells of adult tissues maintain a very low proliferative rate but retain their angiogenic properties, which are rekindled during wound healing[12], tissue regeneration[13], and pathologic processes such as cancer[14], atherosclerosis[15], rheumatoid arthritis[16], diabetic retinopathy[17], and psoriasis[18].

Once formed, neovessels are fated to survive or regress depending on the biologic context in which angiogenesis takes place. Whereas the mechanisms responsible for the initiation of angiogenesis have been extensively investigated, those regulating the reabsorption of neovessels have received less attention and only recently have become the focus of mechanistic studies, as the vascular regression process has been recognized a key determinant of angiogenic outcomes. Understanding how neovessels regress may lead to new pharmacological approaches to treat angiogenesis-dependent disorders. On the other hand, learning what makes neovessels mature and become stable may help design novel strategies for the stable revascularization of ischemic organs and the optimized delivery of anti-neoplastic drugs. This chapter focuses on postangiogenic vascular regression and survival, and briefly reviews selected studies that have led to our current understanding of the mechanisms regulating the late stages of the angiogenic process.

## 2. THE MORPHOLOGY OF VASCULAR REGRESSION

The first detailed observations of vascular regression can be traced back to the early 20$^{th}$ century when Sandison, using a transparent ear chamber that allowed direct observation of angiogenesis in the rabbit, reported that some neovessels of healing wounds differentiated into capillaries, arteries, and veins whereas others retracted or disappeared.[19] Using a modification of the same model, Clark et al. noticed a close relation between blood flow and neovessel survival, and reported that capillaries receiving the greatest blood flow developed into larger and more stable channels, whereas hypoperfused capillaries broke their connections with other capillaries retracting into the parent vessels, or separated altogether from the vascular network, growing shorter and eventually disappearing.[20] In the 1970s Ausprunk et al. reported that capillary regression in the cornea model of angiogenesis was preceded by blood stasis, and suggested that ischemia was responsible for endothelial injury and subsequent influx of scavenging macrophages which removed vascular debris from retracting capillaries.[21] Modlich et al. identified two major mechanisms of blood vessel regression in the corpus luteum: a) detachment of endothelial cells from their basement membrane, followed by shedding of dying cells into the circulation, and b) contraction and occlusion of arterioles and small arteries associated with extensive proliferation of smooth muscle cells.[22] Endothelial detachment in this study was preceded by dissociation of tight junctions and dissolution of the underlying basement membrane; remnants of regressed microvessels lacked an endothelial lining and were composed of smooth muscle cells only, focally embedded in multiple layers of disorganized basement membrane. Ingber et al. observed a similar mechanism of vessel reabsorption in chicken chorioallantoic membranes treated with angiostatic steroids, which induced basement membrane breakdown.[23] Vascular regression in the aortic ring model was similarly characterized by disengagement of endothelial cells from the extracellular matrix, fragmentation of neovessels, retraction of vascular stumps, and reabsorption of isolated neovessel remnants.[24,25] Detachment and shedding of endothelial cells was identified as a mechanism of vascular regression also during the reabsorption of granulation tissue.[26]

## 3. ROLE OF HEMODYNAMIC MECHANISMS IN VASCULAR REGRESSION

The observation that neovessel breakdown is preceded by cessation of blood flow[20] implicates hemodynamic events as major regulators of the vascular regression process. The mechanical shear stress to which endothelial cells are exposed during laminar blood flow has been shown to be a potent inhibitor of endothelial cell apoptosis.[27] Whenever blood flow is perturbed and becomes turbulent, as in arterial vessels narrowed by atheroclerotic plaques, the hemodynamic signals that promote endothelial survival are switched off and the rate of endothelial apoptosis increases.[28] Similarly, a decrease of blood flow in rabbit carotid arteries causes loss of endothelial cells within days of flow reduction.[29] It is therefore not surprising that the angiogenic response of the vessel wall in three-dimensional *ex vivo* models such as the aortic ring assay, which do not have an active blood flow, is regularly followed by vascular regression.[30]

The molecular mechanisms by which blood flow and hemodynamic changes influence neovessel survival or regression are incompletely understood. There is however, evidence that endothelial cells have mechanosensors on their surface, which transduce the tangential component of hemodynamic forces into chemical signals.[31] These molecules include integrins, receptor tyrosine kinases, ion channels, and G proteins.[32,33] Mechanochemical signal transduction involves mitogen-activated protein kinases, protein kinase C, and PI3K/Akt, which activate a number of transcription factors involved in the regulation of apoptosis and cell proliferation.[32,34,35] Physiologic shear stress stimulates the expression of integrins[33], oxygen radical scavenging enzymes[36], and nitric oxide synthase.[37] Endothelial cell-derived nitric oxide maintains a vasodilator tone that is essential for the regulation of blood flow and pressure. Reduction in shear stress below physiologic levels influences the expression of matrix metalloproteinase (MMP)-2, MMP-9 and membrane-type 1 MMP (MT1-MMP), which can collectively digest basement membrane and interstitial matrices.[38-40] Thus, changes in blood flow can dramatically alter cell-matrix interactions, and greatly influence the adhesion properties and survival ability of endothelial and mural cells.

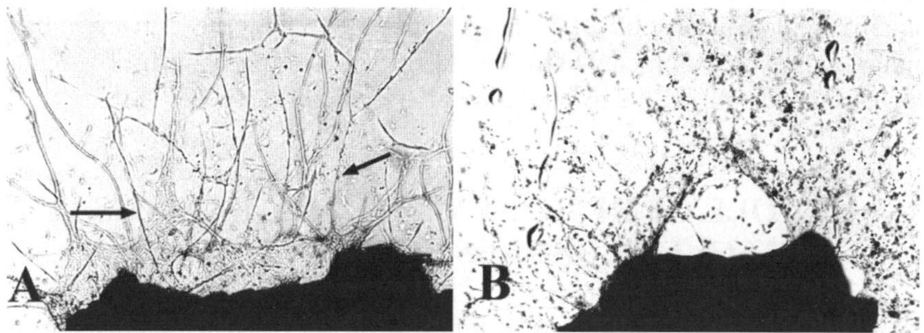


*Figure 4-1.* Survival of angiogenic outgrowths requires vascular endothelial growth factor (VEGF). Explants of rat vena cava cultured in collagen and kept in serum-free medium containing VEGF (20 ng/ml) generated a florid angiogenic response in nine days. A representative VEGF-treated culture shows a complex network of branching microvessels (A, arrows). Removal of VEGF in a parallel previously vascularized culture caused extensive disintegration of neovessels and massive apoptosis within 24 hours (B). This experiment demonstrates how neovessels are strictly dependent on their angiogenic stimuli and exquisitively sensitive to angiogenic factor withdrawal. Magnification 28X.

## 4. SURVIVAL ROLE OF ANGIOGENIC GROWTH FACTORS

Formation of new blood vessels during angiogenesis is strictly dependent on the activity of soluble angiogenic factors, which stimulate endothelial migration, proliferation, and proteolytic activity.[41] Angiogenic factors play a critical role in the induction of the angiogenic process but are also involved in the regulation of neovessel survival. For example, VEGF is expressed in the corpus luteum through most of the ovarian cycle but is downregulated during luteolysis.[8,42] Angiogenesis in the rat aortic ring assay and in the injured rabbit cornea is regulated by endogenous VEGF, and reduction in VEGF levels during the late stages of the angiogenic process coincides with vascular regression in both models.[43,44] Direct evidence in favor of a survival role for angiogenic factors was demonstrated in the cornea model where withdrawal of the angiogenic stimulus resulted in vascular regression.[21] In an *in vitro* model of bFGF-induced capillary tube formation, bFGF withdrawal led to endothelial apoptosis and disruption of the tube structures.[45] Similar results were obtained in VEGF-treated angiogenic cultures of vena cava, where VEGF withdrawal resulted in massive apoptosis and vascular regression (Figure 4-1). Treatment of experimental tumors in mice with anti-VEGF antibody or VEGF-trap, a soluble VEGF receptor with anti-VEGF activity, caused regression of the tumor vasculature.[46,47] Using a tetracycline-regulated VEGF expression system in transplanted glioma cells, Benjamin and Keshet demonstrated that VEGF withdrawal caused detachment of endothelial cells and regression of tumor vessels.[48] More recently, McDonald et al. reported that tumor blood vessels that had regressed following treatment with inhibitors of VEGF signaling left behind a ghost-like record of their pre-treatment number in the form of empty sleeves of basement membrane.[49]

The balance of different angiogenic factors is believed to play a critical role in determining the fate of neovessels. A cogent example is provided by the angiopoietins, which are critical regulators of embryonal and postnatal angiogenesis.[50] Ang-1 binds to and activates the tyrosine kinase receptor Tie2, whereas Ang-2 binds to the same receptor without activating it, and is believed to function as a Tie2 antagonist. The Ang-1/Tie2 system promotes endothelial survival and neovessel stabilization through the AKT survival pathway.[51] Alteration of the Ang-1/Ang-2 balance due to overexpression of Ang-2 over Ang-1 results in vascular regression, as demonstrated in the involuting corpus luteum.[8,52] Early upregulation of Ang-2 in brain microvessels co-opted by tumor cells destabilizes endothelial cells, making them susceptible to angiogenic stimulation by VEGF, but in the absence of a coexisting VEGF stimulus overexpression of Ang-2 becomes an apoptotic signal, causing vascular regression.[53]

## 5. STABILIZING EFFECT OF EXTRACELLULAR MATRIX AND CELL ADHESION MOLECULES

The extracellular matrix (ECM) provides structural and stabilizing signals to angiogenic neovessels. Endothelial disengagement from the ECM has been described as a component of the vascular regression process during physiologic and pathologic types of angiogenesis.[23,25,26,54] Endothelial cells bind to the ECM through integrins located on the cell surface. Through their ability to mediate cell-matrix interactions, integrins control cell motility, cell proliferation, cell shape, cytoskeletal organization, protein phosphorylation, and gene transcription.[55] Disruption of cell-matrix interactions through genetic or pharmacologic targeting of integrin receptors leads to disengagement of the endothelium from the ECM with resulting "anoikis", a form of apoptosis due to cell detachment from matrix substrates.[56] Developing neovessels have a characteristically tenuous ECM scaffold, and are particularly susceptible to the action of molecules that target integrin function. For example, pharmacologic targeting of alphavbeta3 integrin, a promiscuous receptor for ECM molecules with an exposed Arg-Gly-Asp (RGD) sequence (fibronectin, vitronectin, and denatured collagen) induces endothelial apoptosis in newly formed blood vessels with resulting disruption of blood vessels and tumor regression.[57] Similarly, synthetic RGD-containing peptides, which compete with the natural ECM ligands for integrin receptor binding, have the capacity to inhibit angiogenesis and induce vascular regression.[58-60] Conversely, neovessels formed in gels of interstitial collagen can be stabilized by incorporating basement membrane molecules into the collagen matrix.[61,62] Proliferating hemangiomas show extensive subendothelial deposition of vitronectin, which is instead absent in regressing hemangiomas.[63] Likewise, capillary regression in the chorion allantoic membrane model following treatment with angiostatic steroids is associated with loss of fibronectin and laminin.[23]

Of comparable importance are cell-cell interactions, which are essential for endothelial cell migration, proliferation and survival during angiogenesis. Detachment of endothelial cells during vascular regression is associated with loss of cell-to-cell contacts, which involve cell adhesion molecules.[22] Among them, vascular endothelial cadherin (VE cadherin) has been shown to play a central role in mediating adhesion between adjacent endothelial cells and regulating their ability to respond to angiogenic stimulation.[64] Disruption of the VE cadherin gene impairs VEGF-mediated endothelial cell survival and angiogenesis[65], and pharmacologic targeting of VE cadherin results in vessel destabilization and regression.[66,67] Cell-cell interactions during capillary morphogenesis and postangiogenic stabilization also involve integrins, which in addition to engaging ECM molecules can also bind to each other[68], and other cell adhesion molecules including platelet-endothelial cell adhesion molecule-1 (PECAM).[69]

## 6. MATRIX METALLOPROTEINASES AS PROMOTERS OF VASCULAR REGRESSION

Upon stimulation by angiogenic factors, endothelial cells of preexisting vessels digest their basement membrane and penetrate the surrounding interstitium using proteolytic enzymes. This invasive process is primarily mediated by MMPs, which can collectively digest all components of the ECM.[70] The discovery that MMPs play a critical role in angiogenesis and cancer spread has led to the development of synthetic inhibitors that chelate zinc atoms required for MMP activity.[71] After initial successes in experimental animal models, this field suffered setbacks as early cancer clinical trials with broad-spectrum MMP inhibitors failed, and side effects were discovered.[70,72] One possible explanation for this unexpected outcome is the multifunctional nature of MMPs, and their involvement not only in angiogenic sprouting but also in postangiogenic vascular regression.

Insights into the role of MMPs in vascular regression were provided by the aortic ring model of angiogenesis.[73] Breakdown of the neovasculature in this assay was found to be associated with progressive accumulation of MMPs and perivascular collagen lysis. Continuous production of MMPs after cessation of angiogenic sprouting was followed by disengagement of endothelial cells from the ECM, apoptosis, fragmentation, retraction, and lysis of the neovasculature. Blockage of MMP function with broad-spectrum inhibitors of MMPs had anti-angiogenic effects if aortic cultures were treated from the beginning of the experiment, but resulted in vascular survival if treatment was initiated after vessels had formed.[25] Similar observations were made in cultures of isolated endothelial cells.[25,74]

*In vivo* studies supported the hypothesis that the vascular regression process was mediated by MMP-related proteolytic events. In fact, elevated levels of MMPs correlated with the involution of the corpus luteum[75] and the lactating mammary gland.[76] A similar pattern of MMP production was demonstrated in the endometrium where MMPs were overexpressed during the late menstrual phase when hemorrhage occurred as a result of vessel breakdown.[77] Interestingly, reduction in blood flow which precedes vascular regression following angiogenesis, has been shown to be associated with upregulated expression of MMP-2 in a rabbit model of arterial injury and repair.[78]

These observations indicate that MMPs have a time- and context-dependent dual role in angiogenesis because they promote neovessel formation during angiogenesis but also mediate vascular regression following angiogenic sprouting. Pharmacologic inhibition of MMPs may therefore have pro-angiogenic or anti-angiogenic effects depending on the developmental stage of a growing neovasculature.

## 7. ENDOGENOUS ANGIOGENIC INHIBITORS AND VASCULAR REGRESSION

The angiogenic process is regulated endogenously by naturally occurring inhibitors that counterbalance the stimulatory action of angiogenic factors.[79] These include interferons, interferon-inducible protein 10, platelet factor-4, thrombospondin-1 and -2, and angiostatic steroids. Many endogenous inhibitors of angiogenesis are present in the extracellular matrix and in the circulation and are produced through a proteolytic process of larger molecules that uncovers antiangiogenic cryptic sites.[79] Some of these factors have the additional capacity to induce vascular regression.

Angiostatin, a cleavage product of plasminogen, is generated by the proteolytic action of macrophage elastase (MMP-12) and other MMPs.[80] Similarly, endostatin, a fragment of collagen XVIII, is produced by elastase activity.[81] Both angiostatin and endostatin cause tumor regression and dormancy through their capacity to induce endothelial apoptosis and inhibit angiogenesis.[82,83] Angiostatin has been shown to induce regression of neovessels in mice with an established corneal neovascularization.[84] Endostatin binds to the alpha5 beta1 integrin and interferes with endothelial attachment to fibronectin.[85] More recently, proteolytic fragments of other extracellular matrix molecules have been shown to have anti-angiogenic activity. For example, tumstatin, produced by MMP-9-mediated proteolysis of type IV collagen, promotes endothelial cell apoptosis and inhibition of angiogenesis by binding to the alphav beta3 integrin receptor.[86] Similarly endorepellin, an inhibitor of angiogenesis derived from the C-terminus of perlecan, causes endothelial cell disassembly of actin cytoskeleton and focal adhesions through alpha2beta1 integrin.[87]

## 8. THE VASCULAR STABILIZING ROLE OF STROMAL CELLS

Endothelial cells are enveloped by mural cells, including pericytes and smooth muscle cells. Mural cells are in turn surrounded by adventitial fibroblasts. Collectively, these nonendothelial mesenchymal cells provide stromal support to the vascular endothelium and produce powerful survival signals for angiogenic neovessels.[88,89] Direct evidence in favor of this important function is provided by *in vitro* vascular bioengineering experiments. Rat aortic endothelial cells cultured on a collagen gel rapidly reoarganize into a capillary network when overlaid with a second layer of collagen.[90] If kept under serum-free conditions this vascular network disintegrates within 3 days through a process of apoptosis and vessel fragmentation. Addition of fibroblasts to the collagen gel stabilizes the microvessels, which remain viable for several weeks.[91] A similar effect can be obtained with aortic intima-derived smooth muscle cells, which transform into pericytes upon exposure to endothelial cells.[92] Soluble factors produced by stromal cells are likely to participate in this process. For example, Ang-1, which is abundantly produced by stromal cells, has been shown to promote the survival of capillary networks in collagen gels.[93,94] Experiments with purified angiogenic factors or media conditioned by the fibroblasts or mural cells however, do not fully reproduce the stabilizing effect of fibroblasts or mural cells.[91] Solid-phase factors such as basement membrane molecules produced in the co-cultures are most likely involved in this process. The juxtacrine effect of stromal cells may involve a reduction in MMP-mediated collagen lysis because it can be partially mimicked with antagonist of MMP function or other protease inhibitors.[25,74] Stabilization of neovessels by mural cells is also regulated by cell adhesion molecules. For example, blocking N-cadherin expression *in vivo* with a neutralizing antibody perturbs adhesion of pericytes to endothelial cells and severely disrupts vessel integrity.[95]

Endothelial cells acquire a mural cell coating during angiogenesis by secreting PDGF-B, which binds to and activates the PDGF receptor-beta in mural cells.[96] Both PDGF-B and the PDGF receptor-beta are required for the maturation of neovessels. Genetic disruption of the PDGF-B or PDGF receptor-beta genes in mouse embryos results in the formation of abnormal vessels without pericytes.[96-98] The lack of pericytes leads to multiple vascular abnormalities including endothelial hyperplasia, increased capillary diameter, abnormal EC shape and ultrastructure, increased transendothelial permeability, and hemorrhages. Similarly, mouse embryos lacking a functioning Ang-1 or its receptor Tie2 suffer from an abnormal vascular development characterized by incomplete vessel maturation and defective assembly of periendothelial mural cells.[99,100] Comparable abnormalities are identified in mutant mouse embryos overexpressing Ang-2, which destabilizes neovessels by acting as a natural Ang-1 antagonist.[101]

The critical importance of mural cells in vascular survival was demonstrated by Keshet and co-workers in an experimental model of tumor angiogenesis in which cancer cells were transfected with a VEGF gene under the control of a tetracycline inducible promoter.[48]

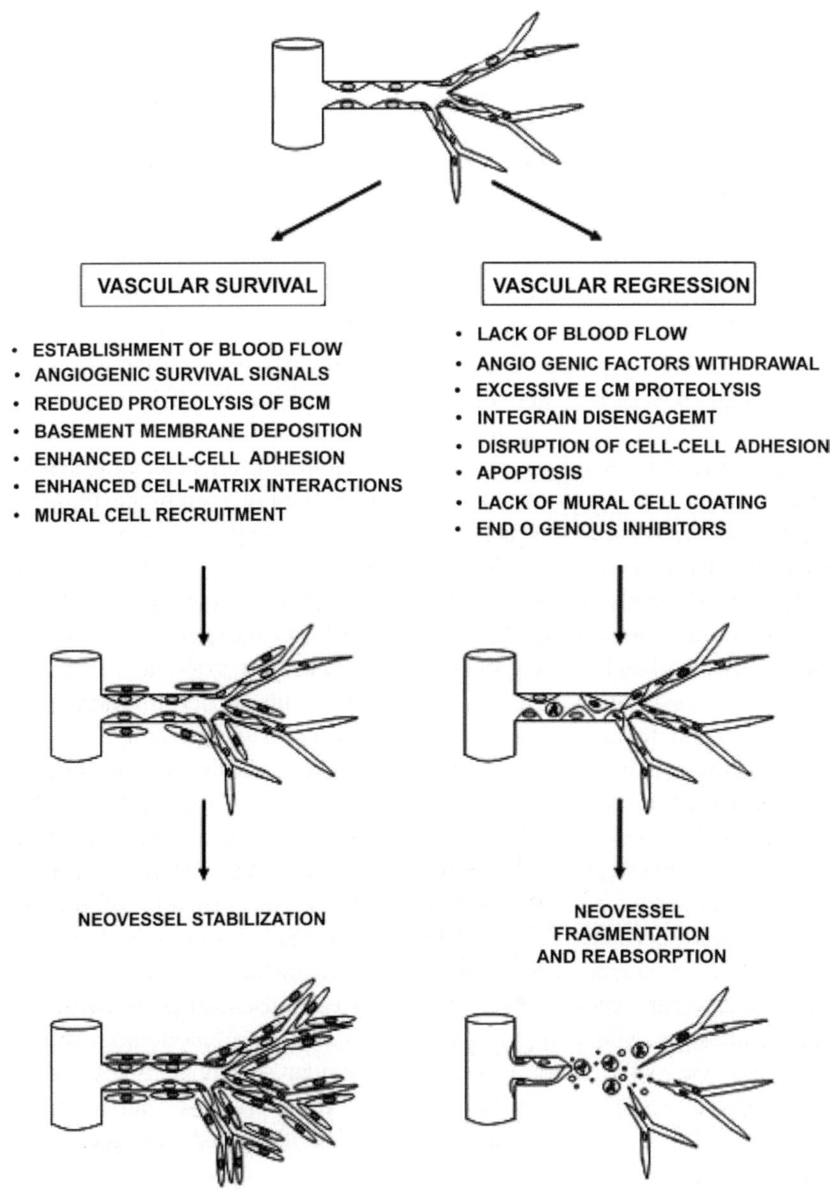


*Figure 4-2.* Mechanisms of vascular survival or regression following angiogenesis.

Pharmacologically induced withdrawal of VEGF production in this model resulted in vascular regression, but neovessel invested by mural cells survived even in the absence of continuous VEGF secretion by tumor cells. Using the RIP-tag model of pancreatic islet tumorigenesis, Hannahan et al. were able to inhibit tumor growth by pharmacologically blocking VEGF receptor function but failed to obtain involution of established tumors with this therapeutic regimen. The same group, however, induced tumor regression by using a combination of drugs that blocked both VEGF and PDGF receptor signaling.[102] Thus, pharmacological targeting of both endothelial cells (VEGF receptor) and pericytes (PDGF receptor) may represent a more effective anti-angiogenic strategy than anti-endothelial therapy alone.

## 9. SUMMARY AND CONCLUSION

Recent advances have moved angiogenesis research from the bench top to the bedside as anti-angiogenic therapy has been recognized a new modality of anti-cancer therapy. The FDA approval of the anti-VEGF drug Avastin for the treatment of colon cancer represents an important milestone in this field[79,103], but more studies are needed to better understand the mechanisms of the angiogenic process and identify angiogenesis-related pathways that can be selectively targeted with specific inhibitors. To that end the recognition that angiogenic outcomes depend not only on the growth of a new vasculature but also on the survival and stabilization of the neovessels has created a new field of research that focuses on the late stages of the angiogenic process. It is well established that the tumor vasculature is chaotic and can be normalized by targeting VEGF and its signaling pathway[104], but more studies are needed to determine what factors regulate the survival of tumor vessels resistant to VEGF inhibition, and to establish whether vascular survival mechanisms in pathologic tissues differ from those operating in the vasculature of normal tissues. The identification in experimental model systems of cellular and molecular mechanisms involved in the regression or stabilization of postangiogenic neovessels (see summary in Figure 4-2) may lead to the development of novel therapeutic approaches to simultaneously target different stages of the angiogenic process, which will hopefully result in a more effective ablation of the tumor vasculature. If successful, these therapies may be applied to other angiogenesis-dependent disorders including diabetic retinopathy, rheumatoid arthritis, psoriasis and hemangiomas. Understanding what makes neovessels survive may also aid in the design of more effective pro-angiogenic strategies to induce formation of stable vessels in patients with coronary artery disease or other ischemic conditions.

# REFERENCES


1. Drake CJ and Little C. The Morphogenesis of Primordial Vascular Networks. In: Vascular Morphogenesis In Vivo, In Vitro, In Mente, edited by Little C, Mironov V and Sage EH. Boston: Birkhauser, 1998, p. 3-19.
2. Wilting J, Kurz H, Oh CW and Christ B. Angiogenesis and Lymphangiogenesis: Analogous Mechanisms and Homologous Growth Factors. In: Vascular Morphogenesis In Vivo, In Vitro, In Mente, edited by Little C, Mironov V and Sage EH. Boston: Birkhauser, 1998, p. 21-34.
3. Caduff JH, Fischer LC and Burri PH. Scanning electron microscope study of the developing microvasculature in the postnatal rat lung. Anat Rec 216: 154-164, 1986.
4. Drake CJ, Hungerford JE and Little CD. Morphogenesis of the first blood vessels. Ann N Y Acad Sci 857: 155-179, 1998.
5. Vrancken Peeters MP, Gittenberger-de Groot AC, Mentink MM, Hungerford JE, Little CD and Poelmann RE. The development of the coronary vessels and their differentiation into arteries and veins in the embryonic quail heart. Dev Dyn 208: 338-348, 1997.
6. Etchevers HC, Vincent C, Le Douarin NM and Couly GF. The cephalic neural crest provides pericytes and smooth muscle cells to all blood vessels of the face and forebrain. Development 128: 1059-1068, 2001.
7. Poelmann RE, Lie-Venema H and Gittenberger-de Groot AC. The role of the epicardium and neural crest as extracardiac contributors to coronary vascular development. Tex Heart Inst J 29: 255-261, 2002.
8. Goede V, Schmidt T, Kimmina S, Kozian D and Augustin HG. Analysis of blood vessel maturation processes during cyclic ovarian angiogenesis. Lab Invest 78: 1385-1394, 1998.
9. Smith SK. Regulation of angiogenesis in the endometrium. Trends Endocrinol Metab 12: 147-151, 2001.
10. Jaffe RB. Importance of angiogenesis in reproductive physiology. Semin Perinatol 24: 79-81, 2000.
11. Djonov V, Andres AC and Ziemiecki A. Vascular remodelling during the normal and malignant life cycle of the mammary gland. Microsc Res Tech 52: 182-189, 2001.
12. Appleton I. Wound healing: future directions. IDrugs 6: 1067-1072, 2003.
13. Furnus CC, Inda AM, Andrini LB, Garcia MN, Garcia AL, Badran AF and Errecalde AL. Chronobiology of the proliferative events related to angiogenesis in mice liver regeneration after partial hepatectomy. Cell Biol Int 27: 383-386, 2003.
14. Folkman J. What is the evidence that tumors are angiogenesis dependent? J Natl Cancer Inst 82: 4-6, 1990.
15. Moulton KS. Plaque angiogenesis and atherosclerosis. Curr Atheroscler Rep 3: 225-233, 2001.
16. Szekanecz Z, Gaspar L and Koch AE. Angiogenesis in rheumatoid arthritis. Front Biosci 10: 1739-1753, 2005.
17. Adamis AP, Aiello LP and D'Amato RA. Angiogenesis and ophthalmic disease. Angiogenesis 3: 9-14, 1999.
18. Arbiser JL. Angiogenesis and the skin: a primer. J Am Acad Dermatol 34: 486-497, 1996.
19. Sandison JC. A new method for the microscopic study of living growing tissues by the introduction of a transparent chamber in the rabbit's ear. Anat Rec 29: 281-290, 1924.
20. Clark ER CE. Microscopic observation on the growth of blood capillaries in the living mammal. Am J Anat 64: 251-299, 1939.

21. Ausprunk DH, Falterman K and Folkman J. The sequence of events in the regression of corneal capillaries. Lab Invest 38: 284-294, 1978.
22. Modlich U, Kaup FJ and Augustin HG. Cyclic angiogenesis and blood vessel regression in the ovary: blood vessel regression during luteolysis involves endothelial cell detachment and vessel occlusion. Lab Invest 74: 771-780, 1996.
23. Ingber DE, Madri JA and Folkman J. A possible mechanism for inhibition of angiogenesis by angiostatic steroids: induction of capillary basement membrane dissolution. Endocrinology 119: 1768-1775, 1986.
24. Nicosia RF. The Rat Aorta Model of Angiogenesis and its Applications. In: Microvascular Morphogenesis in Vivo, in Vitro and in Mente., edited by Mironov V, Little C and Sage H. Boston: Birkhäuser, 1998, p. 111-139.
25. Zhu WH, Guo X, Villaschi S and Nicosia R.F. Regulation of vascular growth and regression by matrix metalloproteinases in the rat aorta model of angiogenesis. Lab Invest 80: 545-555, 2000.
26. Honma T and Hamasaki T. Ultrastructure of blood vessel regression in involution of foreign-body granuloma. J Submicrosc Cytol Pathol 30: 31-44, 1998.
27. Dimmeler S, Haendeler J, Rippmann V, Nehls M and Zeiher AM. Shear stress inhibits apoptosis of human endothelial cells. FEBS Lett 399: 71-74, 1996.
28. Tricot O, Mallat Z, Heymes C, Belmin J, Leseche G and Tedgui A. Relation between endothelial cell apoptosis and blood flow direction in human atherosclerotic plaques. Circulation 101: 2450-2453, 2000.
29. Langille BL, Bendeck MP and Keeley FW. Adaptations of carotid arteries of young and mature rabbits to reduced carotid blood flow. Am J Physiol 256: H931-H939, 1989.
30. Nicosia RF and Ottinetti A. Growth of microvessels in serum-free matrix culture of rat aorta. A quantitative assay of angiogenesis in vitro. Lab Invest 63: 115-122, 1990.
31. Dimmeler S and Zeiher AM. Endothelial cell apoptosis in angiogenesis and vessel regression. Circ Res 87: 434-439, 2000.
32. Kakisis JD, Liapis CD and Sumpio BE. Effects of cyclic strain on vascular cells. Endothelium 11: 17-28, 2004.
33. Chen KD, Li YS, Kim M, Li S, Yuan S, Chien S and Shyy JY. Mechanotransduction in response to shear stress. Roles of receptor tyrosine kinases, integrins, and Shc. J Biol Chem 274: 18393-18400, 1999.
34. Fisher AB, Chien S, Barakat AI and Nerem RM. Endothelial cellular response to altered shear stress. Am J Physiol Lung Cell Mol Physiol 281: L529-L533, 2001.
35. Khachigian LM, Anderson KR, Halnon NJ, Gimbrone MA, Jr., Resnick N and Collins T. Egr-1 is activated in endothelial cells exposed to fluid shear stress and interacts with a novel shear-stress-response element in the PDGF A-chain promoter. Arterioscler Thromb Vasc Biol 17: 2280-2286, 1997.
36. Dimmeler S, Hermann C, Galle J and Zeiher AM. Upregulation of superoxide dismutase and nitric oxide synthase mediates the apoptosis-suppressive effects of shear stress on endothelial cells. Arterioscler Thromb Vasc Biol 19: 656-664, 1999.
37. Cai H, McNally JS, Weber M and Harrison DG. Oscillatory shear stress upregulation of endothelial nitric oxide synthase requires intracellular hydrogen peroxide and CaMKII. J Mol Cell Cardiol 37: 121-125, 2004.
38. Palumbo R, Gaetano C, Melillo G, Toschi E, Remuzzi A and Capogrossi MC. Shear stress downregulation of platelet-derived growth factor receptor-beta and matrix metalloprotease-2 is associated with inhibition of smooth muscle cell invasion and migration. Circulation 102: 225-230, 2000.

39. Magid R, Murphy TJ and Galis ZS. Expression of matrix metalloproteinase-9 in endothelial cells is differentially regulated by shear stress. Role of c-Myc. J Biol Chem 278: 32994-32999, 2003.
40. Yamaguchi S, Yamaguchi M, Yatsuyanagi E, Yun SS, Nakajima N, Madri JA and Sumpio BE. Cyclic strain stimulates early growth response gene product 1-mediated expression of membrane type 1 matrix metalloproteinase in endothelium. Lab Invest 82: 949-956, 2002.
41. Folkman J. Fundamental concepts of the angiogenic process. Curr Mol Med 3: 643-651, 2003.
42. Yamamoto S, Konishi I, Tsuruta Y, Nanbu K, Mandai M, Kuroda H, Matsushita K, Hamid AA, Yura Y and Mori T. Expression of vascular endothelial growth factor (VEGF) during folliculogenesis and corpus luteum formation in the human ovary. Gynecol Endocrinol 11: 371-381, 1997.
43. Nicosia RF, Lin YJ, Hazelton D and Qian X. Endogenous regulation of angiogenesis in the rat aorta model. Role of vascular endothelial growth factor. Am J Pathol 151: 1379-1386, 1997.
44. Edelman JL, Castro MR and Wen Y. Correlation of VEGF expression by leukocytes with the growth and regression of blood vessels in the rat cornea. Invest Ophthalmol Vis Sci 40: 1112-1123, 1999.
45. Satake S, Kuzuya M, Ramos MA, Kanda S and Iguchi A. Angiogenic stimuli are essential for survival of vascular endothelial cells in three-dimensional collagen lattice. Biochem Biophys Res Commun 244: 642-646, 1998.
46. Ferrara N. Role of vascular endothelial growth factor in physiologic and pathologic angiogenesis: therapeutic implications. Semin Oncol 29: 10-14, 2002.
47. Holash J, Davis S, Papadopoulos N, Croll SD, Ho L, Russell M, Boland P, Leidich R, Hylton D, Burova E, Ioffe E, Huang T, Radziejewski C, Bailey K, Fandl JP, Daly T, Wiegand SJ, Yancopoulos GD and Rudge JS. VEGF-Trap: a VEGF blocker with potent antitumor effects. Proc Natl Acad Sci U S A 99: 11393-11398, 2002.
48. Benjamin LE, Golijanin D, Itin A, Pode D and Keshet E. Selective ablation of immature blood vessels in established human tumors follows vascular endothelial growth factor withdrawal. J Clin Invest 103: 159-165, 1999.
49. Inai T, Mancuso M, Hashizume H, Baffert F, Haskell A, Baluk P, Hu-Lowe DD, Shalinsky DR, Thurston G, Yancopoulos GD and McDonald DM. Inhibition of vascular endothelial growth factor (VEGF) signaling in cancer causes loss of endothelial fenestrations, regression of tumor vessels, and appearance of basement membrane ghosts. Am J Pathol 165: 35-52, 2004.
50. Davis S and Yancopoulos GD. The angiopoietins: Yin and Yang in angiogenesis. Curr Top Microbiol Immunol 237: 173-185, 1999.
51. Kim I, Kim HG, So JN, Kim JH, Kwak HJ and Koh GY. Angiopoietin-1 regulates endothelial cell survival through the phosphatidylinositol 3'-Kinase/Akt signal transduction pathway. Circ Res 86: 24-29, 2000.
52. Holash J, Wiegand SJ and Yancopoulos GD. New model of tumor angiogenesis: dynamic balance between vessel regression and growth mediated by angiopoietins and VEGF. Oncogene 18: 5356-5362, 1999.
53. Holash J, Maisonpierre PC, Compton D, Boland P, Alexander CR, Zagzag D, Yancopoulos GD and Wiegand SJ. Vessel cooption, regression, and growth in tumors mediated by angiopoietins and VEGF. Science 284: 1994-1998, 1999.
54. Augustin HG, Braun K, Telemenakis I, Modlich U and Kuhn W. Ovarian angiogenesis. Phenotypic characterization of endothelial cells in a physiological model of blood vessel growth and regression. Am J Pathol 147: 339-351, 1995.

55. Hynes RO. Integrins: bidirectional, allosteric signaling machines. Cell 110: 673-687, 2002.
56. Frisch SM and Ruoslahti E. Integrins and anoikis. Curr Opin Cell Biol 9: 701-706, 1997.
57. Brooks PC, Clark RA and Cheresh DA. Requirement of vascular integrin alpha v beta 3 for angiogenesis. Science 264: 569-571, 1994.
58. Nicosia RF and Bonanno E. Inhibition of angiogenesis in vitro by Arg-Gly-Asp-containing synthetic peptide. Am J Pathol 138: 829-833, 1991.
59. Nicosia RF, Bonanno E and Smith M. Fibronectin promotes the elongation of microvessels during angiogenesis in vitro. J Cell Physiol 154: 654-661, 1993.
60. Storgard CM, Stupack DG, Jonczyk A, Goodman SL, Fox RI and Cheresh DA. Decreased angiogenesis and arthritic disease in rabbits treated with an alphavbeta3 antagonist. J Clin Invest 103: 47-54, 1999.
61. Nicosia RF, Bonanno E, Smith M and Yurchenco P. Modulation of angiogenesis in vitro by laminin-entactin complex. Dev Biol 164: 197-206, 1994.
62. Bonanno E, Iurlaro M, Madri JA and Nicosia RF. Type IV collagen modulates angiogenesis and neovessel survival in the rat aorta model. In Vitro Cell Dev Biol Anim 36: 336-340, 2000.
63. Jang YC, Arumugam S, Ferguson M, Gibran NS and Isik FF. Changes in matrix composition during the growth and regression of human hemangiomas. J Surg Res 80: 9-15, 1998.
64. Dejana E, Bazzoni G and Lampugnani MG. Vascular endothelial (VE)-cadherin: only an intercellular glue? Exp Cell Res 252: 13-19, 1999.
65. Carmeliet P, Lampugnani MG, Moons L, Breviario F, Compernolle V, Bono F, Balconi G, Spagnuolo R, Oostuyse B, Dewerchin M, Zanetti A, Angellilo A, Mattot V, Nuyens D, Lutgens E, Clotman F, de Ruiter MC, Gittenberger-de Groot A, Poelmann R, Lupu F, Herbert JM, Collen D and Dejana E. Targeted deficiency or cytosolic truncation of the VE-cadherin gene in mice impairs VEGF-mediated endothelial survival and angiogenesis. Cell 98: 147-157, 1999.
66. Crosby CV, Fleming PA, Argraves WS, Corada M, Zanetta L, Dejana E and Drake CJ. VE-cadherin is not required for the formation of nascent blood vessels but acts to prevent their disassembly. Blood 105: 2771-2776, 2005.
67. Bach TL, Barsigian C, Chalupowicz DG, Busler D, Yaen CH, Grant DS and Martinez J. VE-Cadherin mediates endothelial cell capillary tube formation in fibrin and collagen gels. Exp Cell Res 238: 324-334, 1998.
68. Dejana E. Endothelial cell adhesive receptors. J Cardiovasc Pharmacol 21 Suppl 1: S18-S21, 1993.
69. DeLisser HM, Christofidou-Solomidou M, Strieter RM, Burdick MD, Robinson CS, Wexler RS, Kerr JS, Garlanda C, Merwin JR, Madri JA and Albelda SM. Involvement of endothelial PECAM-1/CD31 in angiogenesis. Am J Pathol 151: 671-677, 1997.
70. Stetler-Stevenson WG. Matrix metalloproteinases in angiogenesis: a moving target for therapeutic intervention. J Clin Invest 103: 1237-1241, 1999.
71. Borkakoti N. Matrix metalloprotease inhibitors: design from structure. Biochem Soc Trans 32: 17-20, 2004.
72. Ramnath N and Creaven PJ. Matrix metalloproteinase inhibitors. Curr Oncol Rep 6: 96-102, 2004.
73. Nicosia R and Zhu WH. Rat Aortic Ring Assay of Angiogenesis. In: Methods in Endothelial Cell Biology, edited by Augustin H. Berlin Heidelberg: Springer-Verlag, 2004, p. 125-144.

74. Davis GE, Pintar Allen KA, Salazar R and Maxwell SA. Matrix metalloproteinase-1 and -9 activation by plasmin regulates a novel endothelial cell-mediated mechanism of collagen gel contraction and capillary tube regression in three-dimensional collagen matrices. J Cell Sci 114: 917-930, 2001.
75. Duncan WC, McNeilly AS and Illingworth PJ. The effect of luteal "rescue" on the expression and localization of matrix metalloproteinases and their tissue inhibitors in the human corpus luteum. J Clin Endocrinol Metab 83: 2470-2478, 1998.
76. Ambili M, Jayasree K and Sudhakaran PR. 60K gelatinase involved in mammary gland involution is regulated by beta-oestradiol. Biochim Biophys Acta 1403: 219-231, 1998.
77. Salamonsen LA. Matrix metalloproteinases and endometrial remodelling. Cell Biol Int 18: 1139-1144, 1994.
78. Bassiouny HS, Song RH, Hong XF, Singh A, Kocharyan H and Glagov S. Flow regulation of 72-kD collagenase IV (MMP-2) after experimental arterial injury. Circulation 98: 157-163, 1998.
79. Folkman J. Endogenous angiogenesis inhibitors. APMIS 112: 496-507, 2004.
80. Dong Z, Kumar R, Yang X and Fidler IJ. Macrophage-derived metalloelastase is responsible for the generation of angiostatin in Lewis lung carcinoma. Cell 88: 801-810, 1997.
81. Wen W, Moses MA, Wiederschain D, Arbiser JL and Folkman J. The generation of endostatin is mediated by elastase. Cancer Res 59: 6052-6056, 1999.
82. O'Reilly MS, Holmgren L, Chen C and Folkman J. Angiostatin induces and sustains dormancy of human primary tumors in mice. Nat Med 2: 689-692, 1996.
83. O'Reilly MS, Boehm T, Shing Y, Fukai N, Vasios G, Lane WS, Flynn E, Birkhead JR, Olsen BR and Folkman J. Endostatin: an endogenous inhibitor of angiogenesis and tumor growth. Cell 88: 277-285, 1997.
84. Ambati BK, Joussen AM, Ambati J, Moromizato Y, Guha C, Javaherian K, Gillies S, O'Reilly MS and Adamis AP. Angiostatin inhibits and regresses corneal neovascularization. Arch Ophthalmol 120: 1063-1068, 2002.
85. Wickstrom SA, Alitalo K and Keski-Oja J. Endostatin associates with integrin alpha5beta1 and caveolin-1, and activates Src via a tyrosyl phosphatase-dependent pathway in human endothelial cells. Cancer Res 62: 5580-5589, 2002.
86. Sudhakar A, Sugimoto H, Yang C, Lively J, Zeisberg M and Kalluri R. Human tumstatin and human endostatin exhibit distinct antiangiogenic activities mediated by alpha v beta 3 and alpha 5 beta 1 integrins. Proc Natl Acad Sci U S A 100: 4766-4771, 2003.
87. Bix G, Fu J, Gonzalez EM, Macro L, Barker A, Campbell S, Zutter MM, Santoro SA, Kim JK, Hook M, Reed CC and Iozzo RV. Endorepellin causes endothelial cell disassembly of actin cytoskeleton and focal adhesions through alpha2beta1 integrin. J Cell Biol 166: 97-109, 2004.
88. Hirschi KK and D'Amore PA. Control of angiogenesis by the pericyte: molecular mechanisms and significance. EXS 79: 419-428, 1997.
89. Betsholtz C, Lindblom P and Gerhardt H. Role of pericytes in vascular morphogenesis. EXS 115-125, 2005.
90. Montesano R, Orci L and Vassalli P. In vitro rapid organization of endothelial cells into capillary-like networks is promoted by collagen matrices. J Cell Biol 97: 1648-1652, 1983.
91. Villaschi S and Nicosia RF. Paracrine interactions between fibroblasts and endothelial cells in a serum-free coculture model. Modulation of angiogenesis and collagen gel contraction. Lab Invest 71: 291-299, 1994.

92. Nicosia RF and Villaschi S. Rat aortic smooth muscle cells become pericytes during angiogenesis in vitro. Lab Invest 73: 658-666, 1995.
93. Davis S and Yancopoulos GD. The angiopoietins: Yin and Yang in angiogenesis. Curr Top Microbiol Immunol 237: 173-185, 1999.
94. Papapetropoulos A, Garcia-Cardena G, Dengler TJ, Maisonpierre PC, Yancopoulos GD and Sessa WC. Direct actions of angiopoietin-1 on human endothelium: evidence for network stabilization, cell survival, and interaction with other angiogenic growth factors. Lab Invest 79: 213-223, 1999.
95. Gerhardt H, Wolburg H and Redies C. N-cadherin mediates pericytic-endothelial interaction during brain angiogenesis in the chicken. Dev Dyn 218: 472-479, 2000.
96. Hellstrom M, Kal n M, Lindahl P, Abramsson A and Betsholtz C. Role of PDGF-B and PDGFR-beta in recruitment of vascular smooth muscle cells and pericytes during embryonic blood vessel formation in the mouse. Development 126: 3047-3055, 1999.
97. Lindahl P, Johansson BR, Leveen P and Betsholtz C. Pericyte loss and microaneurysm formation in PDGF-B-deficient mice. Science 277: 242-245, 1997.
98. Soriano P. Abnormal kidney development and hematological disorders in PDGF beta-receptor mutant mice. Genes Dev 8: 1888-1896, 1994.
99. Suri C, Jones PF, Patan S, Bartunkova S, Maisonpierre PC, Davis S, Sato TN and Yancopoulos GD. Requisite role of angiopoietin-1, a ligand for the TIE2 receptor, during embryonic angiogenesis. Cell 87: 1171-1180, 1996.
100. Sato TN, Tozawa Y, Deutsch U, Wolburg-Buchholz K, Fujiwara Y, Gendron-Maguire M, Gridley T, Wolburg H, Risau W and Qin Y. Distinct roles of the receptor tyrosine kinases Tie-1 and Tie-2 in blood vessel formation. Nature 376: 70-74, 1995.
101. Maisonpierre PC, Suri C, Jones PF, Bartunkova S, Wiegand SJ, Radziejewski C, Compton D, McClain J, Aldrich TH, Papadopoulos N, Daly TJ, Davis S, Sato TN and Yancopoulos GD. Angiopoietin-2, a natural antagonist for Tie2 that disrupts in vivo angiogenesis. Science 277: 55-60, 1997.
102. Bergers G, Song S, Meyer-Morse N, Bergsland E and Hanahan D. Benefits of targeting both pericytes and endothelial cells in the tumor vasculature with kinase inhibitors. J Clin Invest 111: 1287-1295, 2003.
103. Ferrara N. Napoleone Ferrara discusses Avastin and the future of anti-angiogenesis therapy. Drug Discov Today 10: 539-541, 2005.
104. Jain RK. Tumor angiogenesis and accessibility: role of vascular endothelial growth factor. Semin Oncol 29: 3-9, 2002.

Chapter 5

# VASCULOGENIC MIMICRY: ANGIOGENESIS IN DISGUISE?

Mary J.C. Hendrix, Elisabeth A. Seftor, Richard E.B. Seftor
*Children's Memorial Research Center, Northwestern University Feinberg School of Medicine and Robert H. Lurie Comprehensive Cancer Center*

## 1. VASCULOGENIC MIMICRY

The concept of vasculogenic mimicry (VM) evolved from studies that correlated the molecular profile of aggressive melanoma cells with observations made both *in vitro* using three-dimensional matrices and from histopathological samples from patients with aggressive melanoma. As first defined, VM described the unique characteristic of aggressive melanoma tumor cells to express endothelial-associated genes and form extracellular matrix (ECM)-rich, vasculogenic-like networks in three-dimensional culture. While these networks appeared to recapitulate embryonic vasculogenic networks, they also resembled the distinctive patterned, ECM-rich networks observed in patients' aggressive tumors.[1-11]

Many of the biological properties germane to embryogenesis are also important in tumor growth. During embryonic development, the primary vascular networks are formed during a process called vasculogenesis, the *in situ* differentiation of mesodermal progenitor cells (angioblasts and hemangioblasts) to endothelial cells that organize into a primitive network.[12-14] Angiogenesis is the subsequent remodeling of the vasculogenic network into a more refined microvasculature which results from the sprouting of new capillaries from preexisting networks. In a similar manner during cancer progression, it is generally believed that tumors require a vascular network to serve as a blood supply for growth and as a possible conduit for metastatic dissemination.[15-22]



*R. Forough (ed.), New Frontiers in Angiogenesis*, 97–109.

In aggressive human melanomas and xenograft models, patterned networks lined by tumor cells are observed in the form of loops (and arcs) that appear to encircle spheroidal nests of melanoma. The networks have been shown to be rich in laminin and contain either small channel-like spaces between them, or are partially or totally occluded ECM sheets surrounding nests of tumor cells. Some of the channel-like spaces were originally called vascular channels because they contained red blood cells and plasma and were thought to provide a potential perfusion mechanism and dissemination route within the tumor compartment that functioned either independently of or simultaneously with angiogenesis (or other sources of vascularization, such as vessel cooption). While the presence of VM in a patient's melanoma correlates with a poor prognosis[4-11] and little is known about the biological relevance of this phenomenon, additional studies have reported VM in several other tumor types including breast, lung, prostatic and ovarian carcinoma. These studies are beginning to examine some of the key molecular mechanisms underlying this unique process and in particular, to demonstrate a viable connection of blood flow between tumor cell-lined vascular spaces and endothelial-lined and/or mature vasculature.[23]

Continuing research has focused on the unique differences between highly aggressive and poorly aggressive melanomas in relation to VM and revealed a number of interesting findings in addition to the presence of patterned networks in aggressive melanoma samples. Molecular analyses of over 40 human cutaneous and uveal melanoma cell lines have revealed unusual and unexpected findings regarding the phenotype of aggressive tumors cells.[1,11,24,25] Highly aggressive compared to poorly aggressive melanoma cells (including genetically matched highly and poorly aggressive cell lines from melanoma patients), were analyzed by cDNA microarrays[24] and the spectrum of upregulated genes found to reflect multiple molecular phenotypes, including those associated with progenitor cells, endothelia, epithelia, fibroblasts, hematopoietic lineage, kidney, neuronal lineage, muscle, pericytes, and placental cell types. In addition to the unusual expression of these diverse phenotypes, many of the melanoma-specific genes were down-regulated (excluding the melanoma cell adhesion molecule; MCAM). This molecular profile suggested that aggressive melanoma cells exhibit a deregulated genotype with a genetic footprint reminiscent of an embryonic-like cell with an undifferentiated phenotype and has led to one of the major questions under investigation at this time which concerns the regulation of melanocyte-associated genes. In this regard, MITF (microphthalmia-associated transcription factor) is down-regulated 34-fold in aggressive melanoma cells compared with poorly aggressive tumor cells.[2,25] MITF activates expression of the gene that encodes tyrosinase (an enzyme that is involved in melanocyte differentiation[26]). The tyrosinase gene and tyrosinase-related protein 1 (TYRP1) are also down-regulated 37 to over 100-fold in the same tumor cells (respectively), compared with poorly aggressive tumor cells, and indicates several of the melanocyte-specific genes that are diminished in aggressive melanoma. These results suggest that aggressive melanoma cells could have dedifferentiated *in situ* and may be more difficult to identify with routine histopathology methods.

## 2. MOLECULAR SIGNATURE OF THE MELANOMA VASCULOGENIC PHENOTYPE

With respect to the genes that are upregulated in the aggressive compared to the poorly aggressive cancer cells, many of the genes upregulated in the aggressive cancer cells include those involved in angiogenesis and vasculogenesis; for example, VE (vascular endothelial)-cadherin (CD144 or cadherin 5), EphA2 (erythropoietin-producing hepatocellular carcinoma-A2; epithelial cell kinase), and laminin 5 γ2 chain.[27-29] These molecules (and their binding partners) are required for the formation and maintenance of blood vessels.[12-14,30]

VE-cadherin is an adhesive protein that belongs to the cadherin family of transmembrane proteins, promotes homotypic cell-to-cell interaction[31-34] and prior to the current studies was considered to be endothelial cell-specific. EphA2 is a receptor protein tyrosine kinase that is part of a large family of ephrin receptors[35] that becomes phosphorylated on tyrosine when bound by its ligand ephrin-A1. There is evidence that EphA2 can also be constitutively phosphorylated in unstimulated cells. High expression levels of EphA2 and ephrin-A1 have been associated with increased melanoma thickness, decreased survival and the EphA2/ephrin-A1 pathway has been linked to tumor cell proliferation.[36-38] Laminins are major components of basement membranes and are involved in neurite outgrowth, tumor metastasis, cell attachment and migration, and angiogenesis.[37,39,40] Cleavage of laminin through proteolysis (particularly the laminin 5 γ2 chain), can alter and regulate the integrin-mediated migratory behavior of certain cells[37,41-43] which suggests its potential importance as a molecular trigger in the microenvironment.

At the protein level, VE-cadherin, EphA2, and laminin 5 γ2 chain are expressed by aggressive tumor cells but not by nonaggressive or poorly aggressive melanoma cells.[27-29] The relevance of each protein to the process of VM was demonstrated by down-regulating the expression of these proteins independently and measuring the consequence on the formation of vasculogenic-like networks by aggressive tumor cells *in vitro*. It was found that down-regulating any one of these three proteins (VE-cadherin, EphA2 or laminin 5 γ2 chain) completely abrogated the formation of vasculogenic-like networks in three-dimensional culture by the aggressive melanoma cells.[27-29]

An additional study focused on the tumor cell associated extracellular matrix and its role in VM. This study found that poorly aggressive melanoma cells could be induced to assume a vasculogenic and more migratory phenotype by a microenvironment preconditioned or remodeled by aggressive melanoma cells (subsequent to removal of the aggressive cells[29]). It was shown that a cooperative interaction of laminin 5 γ2 chain with MMP-2 (matrix metalloproteinase-2) and MT1-MMP (membrane type-1 matrix metalloproteinase) to produce laminin 5 γ2' and γ2x promigratory fragments was required for melanoma VM and that laminin co-localized with VM networks both *in vitro* and *in vivo*. These findings indicated that highly aggressive melanoma cells could deposit molecular messages or signals (*i.e.* laminin 5 γ2' and γ2x promigratory fragments) in their microenvironment through the proteolytic activity of MT1-MMP and MMP-2 on laminin 5 γ2 chain and that these interactions are required for the induction of the vasculogenic phenotype in poorly aggressive cells. Furthermore, it was also found that the aggressive cell preconditioned matrix induced the poorly aggressive melanoma cells to express the vascular-associated genes VE-cadherin, EphA2 and laminin 5 γ2 chain. This inductive potential could be "neutralized" by treating the matrix with a potent inhibitor of MMP activity (COL-3, a chemically modified tetracycline; CMT-3) to inhibit MT1-MMP and MMP-2 activity and generation of the laminin 5 γ2' and γ2x promigratory fragments.[44]

There have been a number of different interpretations concerning the concept of VM based on further analyses of the original findings and additional studies that have reported several scenarios related to VM. While an initial interpretation was simply descriptive and defined VM as the patterned networks visible in aggressive tumors when stained using periodic acid Schiff (PAS)-staining, other studies referred to VM as tumor cells that line spaces or channels that contain red blood cells (RBCs) or blood lakes. VM has also been equated to tumor cells expressing endothelial-specific genes. Although it is possible that a combination of any of the above scenarios could be associated with VM in aggressive cancers, the term "vascular mimicry" has also been used synonymously with VM. This term carries broader implications because it could apply to other vascular associated cell types including lymphocytes, macrophages and other molecular phenotypes that could also be expressed by aggressive tumor cells.

## 3. FUNCTIONAL RELEVANCE OF VASCULOGENIC MIMICRY

Studies addressing the functional significance of VM in human melanoma and xenograft models are far more complex than the *in vitro* cellular and molecular investigations and have raised more questions than answers about the relevance of the *in vivo* studies. Two major questions under study regarding melanoma VM are: 1) is there a morphological and functional connection between melanoma tumor cell-lined networks and endothelial-lined vasculature?, and 2) based on their vascular phenotype, is it possible for aggressive melanoma cells to provide a vascular function when challenged to an ischemic, non-tumor microenvironment?

The first question addresses a potential biological and functional connection between tumor cell-lined PAS- and laminin-positive matrix networks within aggressive melanoma tumors and the endothelial-lined vasculature. Studies prior to the introduction of the concept of VM involving aggressive melanoma and other tumors reported that tumor cells could line channels, lakes, sinuses, and come into contact with RBCs.[45-47] However, it was unclear whether these observations were relevant to the functional provision of a blood supply to a growing tumor mass. The prevailing hypothesis at that time was that any blood (RBCs) found in extravascular spaces was probably due to leaky vessels.[48] Furthermore, the PAS-positive patterned networks that were found in aggressive melanoma tumors and correlated with poor clinical outcome[4-11] appeared to morphologically converge with blood vessels.[1,3] This suggested that a putative anastomosis occurred between the tumor cell-lined networks and the endothelial-lined vasculature which would provide a biological source for the occasional RBCs observed within the network infrastructure.[3] This reasoning led to the speculation that the tumor cell-lined networks could provide a unique paracirculation independent of or simultaneous with angiogenesis and/or vessel cooption. While the complex interactions associated with this model will require rigorous scientific scrutiny and examination, a potential tool for these studies is an orthotopic model for human uveal melanoma in SCID mice. This model was developed to study the generation of the unique network patterning characteristic of genetically deregulated aggressive melanoma cells and might provide important new insights into the VM process.[49]

Significant new findings pertinent to addressing the first question have revealed the presence of a fluid-conducting meshwork in xenografts of human cutaneous and uveal melanoma that corresponds to the PAS- and laminin-positive patterned networks.[50,51] Using routine, confocal, and immunoelectron microscopy with a combination of intravenous tracers these studies have shown that tracers may be found inside traditional, endothelial-lined vasculature and extravascularly along channel-like spaces created by PAS/laminin-positive patterned networks surrounding spheroidal nests of tumor cells.[50-52] This surrounding network (described as a fluid conducting meshwork), contained plasma[51] and RBCs (in some instances) which were most likely derived from local tumor vessels that were leaky and undergoing remodeling. While the functional relevance of this finding remains unclear, several theoretical possibilities may exist. The fluid conducting meshwork may provide a nutritional exchange for aggressive tumors to prevent early necrosis, or could be analogous to an edematous inflammatory response where increased pressure leads to the escape of fluid along connective tissue pathways within intra-tissue spaces, or the complex geometry of laminin matrix that is contiguous to the spheroidal nests of cells may act as a suppressive shield against immune surveillance. Nonetheless, the consensus is that the microcirculation of aggressive tumors is complex and may consist of mosaic vessels[53], coopted vessels[54], and angiogenic vessels.[15-22] Additional evidence that there are intratumoral, tumor cell-lined ECM-rich patterned networks capable of providing an extravascular fluid pathway (*i.e.* fluid conducting meshwork[50,51]) suggested that the entire microcirculation in aggressive tumors appears to be a combination of these elements and could result from destructive tumor growth and remodeling.

Recently, injection of fluorescent-tagged aggressive human cutaneous melanoma cells subcutaneously in a nude mouse model permitted a more comprehensive study of blood supply to primary tumour.[55] The mouse vasculature was perfused with a fluorescent tag and microbeads during tumor development and subsequent confocal microscopy revealed the close approximation of tumor cell-lined networks with angiogenic mouse vessels at the human-mouse interface. The delivery of microbeads from the endothelial-lined mouse vasculature to the tumor cell-lined networks suggested a physiological connection between the two compartments. Further destructive growth of melanoma into the vasculature led to the observation of RBCs and plasma (presumably due to leakage) into the tumor cell-lined fluid conducting meshwork. However, the question of whether a direct morphological and physiological anastomosis exists between endothelial-lined vasculature and tumor cell-lined fluid conducting meshwork remains enigmatic. Assessment of physiological blood flow using color Doppler imaging of human melanoma xenografts have shown pulsatile turbulent flow at the mouse-human tissue interface (with mouse endothelial-lined neovasculature) and the central region of the tumor containing melanoma cell-lined networks.[56] If a dynamic, functional exchange of blood through a tumor cell-lined meshwork (rich in laminin) is occurring, this could be related to the observation that aggressive melanoma cells overexpress the anti-coagulant factors TFPI-1 and TFPI-2[56] and appear to exhibit similar anti-coagulant mechanisms as endothelial cells which may contribute to the perfusability of the fluid conducting meshwork. Although these preliminary findings are quiet intriguing, more detailed analyses are required to understand the precise sequence of events associated with the establishment of vascularization during tumor growth and remodeling in highly aggressive *versus* poorly aggressive tumors *versus* normal tissues.

An alternative role for the PAS/laminin-rich fluid conducting meshwork is that it represents an early survival mechanism for nutrient exchange and fluid pressure release which might be replaced by endothelial cells either from nearby angiogenic vessels or from the bone marrow.[57] This intriguing possibility may provide a different perspective on the vasculogenic phenotype of melanoma cells that line and appear to disseminate through these matrix meshworks and will require a higher standard of discrimination to unequivocally identify melanoma cells from endothelial cells.

An additional unexpected finding is that aggressive melanoma cells overexpress VEGF-C, a lymphangiogenesis-associated gene and growth factor for lymphatic vessels. It is also interestingly to note that despite the fact that uveal melanomas lack traditional lymphatic vessels, overexpression of VEGF-C has been found by Clarijs and coworkers.[58] Although lymphangiogenesis often accompanies angiogenesis, the introduction of new research tools[59-61] is providing the means to examine this understudied process. In this regards, Ruoslahti and colleagues have shown localization of LYVE-1 (lymphatic vessel marker) within aggressive cutaneous melanoma[62] which raises an intriguing possibility that the fluid conducting meshwork could "mimic" a lymphatic-like network with circulating fluid and proteins that might leak out of the blood vasculature.

A second question under investigation concerning the biological implications of VM has focused on the plasticity of aggressive melanoma cells in various microenvironments. Aggressive melanoma cells were introduced to an ischemic, non-tumor environment and examinined with respect to their ability to participate in neovascularization and/or to form a tumor mass. In this investigation, fluorescently-labeled human cutaneous metastatic melanoma cells were injected into ischemic hindlimbs of nude mice.[63] The limb vasculature was reperfused after five days and histological cross-sections of newly formed vasculature of reperfused limbs revealed human melanoma cells adjacent to and overlapping with mouse endothelial cells in a linear arrangement forming chimeric vessels – is this angiogenic mimicry? This study demonstrated that the microenvironment could exert a powerful influence on the transendothelial differentiation of malignant melanoma cells involved in neovascularization and reperfusion. Furthermore, Notch proteins which promote the differentiation of endothelial cells into vascular networks[64,65] were examined, and Notch 4 was found to be highly expressed by malignant melanoma cells as they participated in neovascularization. Activated Notch 4 expression has also been detected in patients' invasive melanomas (personal communication, Brian Nickoloff), and interruption of Notch signaling triggers apoptosis of malignant melanoma.[66]

Notch signaling molecules are integrally involved in cell fate determination of stem cells, particularly angioblasts, and nonterminally differentiated cell types.[64,65] Alterations in Notch expression and signaling have also been implicated in human T-cell leukemia[67], murine mammary carcinoma[68], cervical carcinoma[69], prostatic tumour progression[70], and human Ras-transformed cells.[71] Notch signaling molecules in aggressive melanoma cells may offer new clues into the regulation of cell fate decision making pathways that are triggered in the undifferentiated tumor cell phenotype. These observations advance our understanding of the inductive nature of the microenvironment on aggressive tumor cells that express vasculogenic/angiogenic/lymphatic molecules and cell-fate determination proteins associated with a transendothelial phenotype. These findings present new possibilities for therapeutic strategies and novel perspectives on tumor cell plasticity, and emphasize the importance of early differentiation pathways, such as those involved in Notch signaling.[72]

## 4. CLINICAL IMPLICATIONS OF VASCULOGENIC MIMICRY

While a great deal of effort has been devoted to targeting angiogenesis and lymphangiogenesis in cancer patients[15-18,22,30,59-62,73-82], the heterogeneity of the tumor vasculature presents both an opportunity as well as a significant clinical challenge, including issues of drug resistance.[73,78,81,82] Endothelial cells involved in vasculogenesis and angiogenesis are critical for blood vessel development and are key targets in cancer therapy. A recent study from our laboratory addressed whether angiogenesis inhibitors are able to inhibit melanoma cell vasculogenic mimicry in a similar manner as observed in endothelial cells.[83] The data revealed that anginex, TNP-470 and endostatin markedly inhibited cord and tube formation by HMEC-1 and HUVEC endothelial cells *in vitro*; however, tubular networks formed by aggressive MUM-2B and C8161 melanoma cells remained relatively unaffected. Dramatic differences were found in the expression of the putative receptor/signaling complex for endostatin comprised of the integrin $\alpha_5$-subunit and heparin sulfate proteoglycan (perlecan; HSPG2), by endothelial cells *vs.* melanoma, which may provide important clues for the differential response to angiogenesis inhibitors. These findings have the potential to contribute to the development of new agents for anti-vascular therapy – targeting both angiogenesis and tumor cell vasculogenic mimicry.

## REFERENCES


1. Maniotis, A.J. *et al.* Vascular channel formation by human melanoma cells in vivo and in vitro: vasculogenic mimicry. Am. J. Pathol. 155, 739-752 (1999).
2. Seftor, E.A. *et al.* Molecular determinants of human uveal melanoma invasion and metastasis. Clin. Exp. Metastas. 19, 233-246 (2002).
3. Folberg, R., Hendrix, M.J.C. and Maniotis, A.J. Vasculogenic mimicry and tumor angiogenesis. Am. J. Pathol. 156, 361-381 (2000).
4. Folberg, R. *et al.* The prognostic value of tumor blood vessel morphology in primary uveal melanoma. Ophthalmology 100, 1389-1398 (1993).
5. Makitie, T., Summanen, P., Tarkkanen, A. and Kivela, T. Microvascular loops and networks as prognostic indicators in choroidal and ciliary body melanomas. J. Natl. Cancer Inst. 91, 359-367 (1999).
6. Sakamoto, T. *et al.* Histologic findings and prognosis of uveal malignant melanoma in Japanese patients. Am. J. Ophthalmol. 121, 276-283 (1996).
7. Seregard, S., Spangberg, B., Juul, C. and Oskarsson, M. Prognostic accuracy of the mean of the largest nucleoli, vascular patterns, and PC-10 in posterior uveal melanoma. Ophthalmology 105, 485-491 (1998).
8. Thies, A., Mangold, U., Moll, I. and Schumacher, U. PAS-positive loops and networks as a prognostic indicator in cutaneous malignant melanoma. J. Pathol. 195, 537-542 (2001).
9. Warso, M.A. *et al.* Prognostic significance of periodic acid-Schiff-positive patterns in primary cutaneous melanoma. Clin. Cancer Res. 7, 473-477 (2001).
10. Rummelt, V. *et al.* Microcirculation architecture of metastases from primary ciliary body and choroidal melanomas. Am. J. Ophthalmol. 126, 303-305 (1998).
11. Hendrix, M.J.C., Seftor, E.A., Hess, A.R. and Seftor, R.E.B. Vasculogenic mimicry and tumour-cell plasticity: Lessons from melanoma. Nat. Rev Cancer 3:411-421 (2003).
12. Risau, W. Mechanisms of angiogenesis. Nature 386, 671-674 (1997).
13. Carmeliet, P. Mechanisms of angiogenesis and arteriogenesis. Nature Med. 6, 389-395 (2000).
14. Assembly of the Vasculature and Its Regulation. R.J. Tomanek, ed. Birkhauser, Boston, (2002).
15. Folkman, J. Seminars in Medicine of the Beth Israel Hospital, Boston. Clinical applications of research on angiogenesis. New Eng. J. Med. 333, 1757-1763 (1995).
16. Rak, J. and Kerbel, R.S. Treating cancer by inhibiting angiogenesis: new hopes and potential pitfalls. Cancer Metastas. Rev. 15, 231-236 (1996).
17. Kumar, R. and Fidler, I.J. Angiogenic molecules and cancer metastasis. In Vivo 12, 27-34 (1998).
18. Carmeliet, P. and Jain, R.K. Angiogenesis in cancer and other diseases. Nature 407, 249-257 (2000).
19. Gullino, P.M. Angiogenesis and oncogenesis. J. Natl. Cancer Inst. 61, 639-643 (1978).
20. Hanahan, D. and Weinberg, R.A. The hallmarks of cancer. Cell 100, 57-70 (2000).
21. Bouck, N., Stellmach, V. and Hsu, S.C. How tumors become angiogenic. Adv. Cancer Res. 69, 135-174 (1996).

22. Kerbel, R.S. Tumor angiogenesis: past, present and the near future. Carcinogenesis 21, 505-515 (2000).
23. Shirakawa, K. *et al.* Hemodynamics in vasculogenic mimicry and angiogenesis of inflammatory breast cancer xenograft. Cancer Res. 62, 560-566 (2002).
24. Bittner, M. *et al.* Molecular classification of cutaneous malignant melanoma by gene expression profiling. Nature 406, 536-540 (2000).
25. Seftor, E.A. *et al.* Expression of multiple molecular phenotypes by aggressive melanoma tumor cells: role in vasculogenic mimicry. Crit. Rev. Oncol. Hematol. 44, 17-27 (2002).
26. Tachibana, M. *et al.* Ectopic expression of MITF, a gene for Waardenburg syndrome type 2, converts fibroblasts to cells with melanocytes characteristics. Nature Genet. 14, 50-54 (1996).
27. Hess, A.R. *et al.* Molecular regulation of tumor cell vasculogenic mimicry by tyrosine phosphorylation: role of epithelial cell kinase (Eck/EphA2). Cancer Res. 61, 3250-3255 (2001).
28. Hendrix, M.J.C. *et al.* Expression and functional significance of VE-cadherin in aggressive human melanoma cells: role in vasculogenic mimicry. Proc. Natl. Acad. Sci., USA 98, 8018-8023 (2001).
29. Seftor, R.E.B. *et al.* Cooperative interactions of laminin 5 $\gamma 2$ chain, matrix metalloproteinase-2, and membrane type-1-matrix/metalloproteinase are required for mimicry of embryonic vasculogenesis by aggressive melanoma. Cancer Res. 61, 6322-6327 (2001).
30. Hynes, R.O., Bader, B.L. and Hodivala-Diike, K. Integrins in vascular development. Braz. J. Med. Biol. Res. 32, 501-510 (1999).
31. Hynes, R.O. Specificity of cell adhesion in development: the cadherin superfamily. Curr. Opin. Genet. Dev. 2, 621-624 (1992).
32. Kemler, R. Classical cadherins. Semin. Cell Biol. 3, 149-155 (1992).
33. Lampugnani, M.G. A novel endothelial-specific membrane protein is a marker of cell-cell contacts. J. Cell Biol. 118, 1511-1522 (1992).
34. Gumbiner, B.M. Cell adhesion: The molecular basis of tissue architecture and morphogenesis. Cell 4, 345-357 (1996).
35. Pasquale, E.B. The Eph family of receptors. Curr. Opin. Cell Biol. 9, 608-615 (1997).
36. Rosenburg, I.M., Goke, M., Kanai, M., Reinecker, H.C. and Podolsky, D.K. Epithelial cell kinase-B-61: an autocrine loop modulating intestinal epithelial migration and barrier function. Am. J. Physiol. 273(4 part 1), G824-G832 (1997).
37. Straume, O. and Akslen, L.A. Importance of vascular phenotype by basic fibroblast growth factor, and influence of the angiogenic factors basic fibroblast growth factor/fibroblast growth factor receptor-1 and Ephrin-A1/EphA2 on melanoma progression. Am. J. Pathol. 160, 1009-1019 (2001).
38. Easty, D.J. *et al.* Up-regulation of ephrin-A1 during melanoma progression. Int. J. Cancer 84, 494-501 (1999).
39. Malinda, K.M., and Kleinman, H.K. The laminins. Int. J. Biochem. Cell Biol. 289, 57-95 (1996).
40. Colognato, H. and Yurchenco, P.D. Form and function: the laminin family of heterotrimers. Dev. Dynamics 218, 213-234 (2000).
41. Malinda, K.M. *et al.* Identification of laminin $\alpha 1$ and $\beta 1$ chain peptides active for endothelial cell adhesion, tube formation, and aortic sprouting. FASEB J. 13, 53-62 (1999).

42. Koshikawa, N., Giannelli, G., Cirulli, V., Miyazaki, K. and Quaranta, V. Role of cell surface metalloprotease MT1-MMP in epithelial cell migration over laminin-5. J. Cell Biol. 148, 615-624 (2000).
43. Giannelli, G., Falk-Marzillier, J., Schiraldi, O., Stetler-Stevenson, W.G. and Quaranta, V. Induction of cell migration by matrix metalloprotease-2 cleavage of laminin-5. Science 277, 225-228 (1997).
44. Seftor, R.E.B., Seftor, E.A., Kirschmann, D.A. and Hendrix, M.J.C. Targeting the tumor microenvironment with chemically modified tetracyclines: inhibition of laminin 5 $\gamma 2$ chain promigratory fragments and vasculogenic mimicry. Mol. Cancer Therapeut. 1, 1173-1179 (2002).
45. Shubik, P. and Warren, B.A. Additional literature on "vasculogenic mimicry" not cited. Am. J. Pathol. 156, 736 (2000).
46. Warren, B.A. and Shubik, P. The growth of the blood supply to melanoma transplants in the hamster cheek pouch. Lab. Invest. 15, 464-478 (1966).
47. Tímár, J. and Tóth, J. Tumor sinuses-vascular channels. Pathol. Oncol. Res. 6, 83-86 Oncol Res. 2000;6(2):83-6.
48. Hashizuma, H. *et al.* Openings between defective endothelial cells explain tumor vessel leakiness. Am. J. Pathol. 156, 1363-1380 (2000).
49. Mueller, A.J. *et al.* An orthotopic model for human uveal melanoma in SCID mice. Microvasc. Res. 64, 207-213 (2002).
50. Clarijs, R., Otte-Holler, I., Ruiter, D.J. and de Waal, R.M.W. Presence of a fluid-conducting meshwork in xenografted cutaneous and primary human uveal melanoma. Invest. Ophthalmol. Vis. Sci. 43, 912-918 (2002).
51. Maniotis, A.J. *et al.* Control of melanoma morphogenesis, endothelial survival, and perfusion by extracellular matrix. Lab. Invest. 82, 1031-1043 (2002).
52. Potgens, A.J.G., van Altena, M.C., Lubsen, N.H., Ruiter, D.J. and de Waal, R.M.W. Analysis of the tumor vasculature and metastatic behavior of xenografts of human melanoma cell lines transfected with vascular permeability factor. Am. J. Pathol. 148, 1203-1217 (1996).
53. Chang, Y.S., di Tomaso, E., McDonald, D.M., Jones, R., Jain, R.K. and Munn, L.L. Mosaic blood vessels in tumors: frequency of cancer cells in contact with flowing blood. Proc. Natl. Acad. Sci., USA 97, 14608-14613 (2000).
54. Döme, B., Paku, S., Somlai, B. and Tímár, J. Vascularization of cutaneous melanoma involves vessel co-option and has clinical significance. J. Pathol. 197, 355-362 (2002).
55. Goetz, D.E., Yu, J.L., Kerbel, R.s., Burns, P.N. and Foster, F.S. High-frequency Doppler ultrasound monitors the effects of antivascular therapy on tumor blood flow. Cancer Res. 62, 6371-6375 (2002).
56. Ruf, W. *et al.* Differential role of tissue factor pathway inhibitors 1 and 2 in melanoma vasculogenic mimicry. Cancer Res. 63:5381-5389 (2003).
57. Hattori, H. *et al.* Placental growth factor reconstitutes hematopoiesis by recruiting VEGFR1$^+$ stem cells from bone-marrow microenvironment. Nature Med. 8, 841-849 (2002).
58. Clarijs, R., Schalkwijk, L., Ruiter, D.J. and de Waal, R.M.W. Lack of lymph-angiogenesis despite coexpression of VEGF-C and its receptor Flt-4 in uveal melanoma. Invest. Ophthalmol. Vis. Sci. 42, 1422-1428 (2001).
59. Witte, M.H., Bernas, M.J., Martin, C.P. and Witte, C.L. Lymphangiogenesis and lymph-angiodysplasia: from molecular to clinical lymphology. Microscopy Res. Tech. 55, 122-145 (2001).
60. Alitalo, K. and Carmeliet, P. Molecular mechanisms of lymphangiogenesis in health and disease. Cancer Cell 1, 219-227 (2002).

61. Padera, T.P. *et al.* Lymphatic metastasis in the absence of functional intratumor lymphatics. Science 296, 1883-1886 (2002).
62. Laakkonen, P, Porkka, K., Hoffman, J.A. and Ruoslahti, E. A tumor-homing peptide with a targeting specificity related to lymphatic vessels. Nature Med. 8, 751-755 (2002).
63. Hendrix, M.J.C. *et al.* Transendothelial function of human metastatic melanoma cells: role of the microenvironment in cell-fate determination. Cancer Res. 62, 665-668 (2002).
64. Uyttendaele H., Ho, J., Rossant, J. and Kitajewski, J. Vascular patterning defects associated with expression of activated Notch4 in embryonic endothelium. Proc. Natl. Acad. Sci., USA 98, 5643-5648 (2001).
65. Gridley, T. Notch signaling during vascular development. Proc. Natl. Acad. Sci., USA 98,10733-10738 (2001).
66. Quin, J.-Z. *et al.* Interrupting activated Notch signaling triggers apoptosis in melanoma cells. Proc. Soc. Invest. Dermatol., in press (2003).
67. Ellisen, L.W. *et al.* TAN-1, the human homolog of the Drosophila notch gene, is broken by chromosomal translocations in T lymphoblastic neoplasms. Cell 66, 649-661 (1991).
68. Robbins, J., Blonel, B.J., Gallahan, D. and Callahan, R. Mouse mammary tumor gene Int-3: A member of the Notch gene family transforms mammary epithelial cells. J. Virol. 66, 2594-2599 (1992).
69. Zagouras, P., Stifani, S., Blaumueller, C.M., Carcangiu, M.L. and Artavanis-Tsakonas, S. Alterations in Notch signaling in neoplastic lesions of the human cervix. Proc. Natl. Acad. Sci., USA 92, 6414-6418 (1995).
70. Shou, J., Ross, S., Koeppen, H., de Sauvage, F.J. and Gao, W.-Q. Dynamics of Notch expression during murine prostate development and tumorigenesis. Cancer Res. 61, 7291-7297 (2001).
71. Weijzen, S. *et al.* Activation of Notch-1 signaling maintains the neoplastic phenotype in human Ras-transformed cells. Nature Med. 8, 979-986 (2002).
72. Allenspach, E.J., Maillard, I., Aster, J.C. and Pear, W.S. Notch signaling in cancer. Cancer Biol. Therapy 1, 466-476 (2002).
73. Pasqualini, R., Arap, W. and McDonald, D.M. Probing the structural and molecular diversity of tumor vasculature. Trends Mol. Med. 8, 563-571 (2002).
74. Jain, R.K. Normalizing tumor vasculature with anti-angiogenic therapy: a new paradigm for combination therapy. Nature Med. 7, 987-989 (2001).
75. O'Reilly, M.S. Vessel maneuvers: vaccine targets tumor vasculature. Nature Med. 8, 1352-1354 (2002).
76. Miller, K.D., Sweeney, C.J. and Sledge, G.W. Jr. The snark is a boojum: the continuing problem of drug resistance in the antiangiogenic era. Annals Oncol. 14, 20-28 (2003).
77. Scappaticci, F.A. Mechanisms and future directions for angiogenesis-based cancer therapies. J. Clin. Oncol. 20, 3906-3927 (2002).
78. Gee, M.S. *et al.* Tumor vessel development and maturation impose limits on the effectiveness of anti-vascular therapy. Am. J. Pathol. 162, 183-193 (2003).
79. Grossman, D. and Altieri, D.C. Drug resistance in melanoma: mechanisms, apoptosis, and new potential therapeutic targets. Cancer Metast. Rev. 20, 3-11 (2001).
80. Eberhard A, Kahlert S, Goede V, Hemmerlein B, Plate KH, Augustin HG. Heterogeneity of angiogenesis and blood vessel maturation in human tumors: implications for antiangiogenic tumor therapies. Cancer Res 2000;60:1388-3.
81. Coussens, L.M., Fingleton, B. and Matrisian, L.M. Matrix metalloproteinase inhibitors and cancer: trials and tribulations. Science 295, 2387-2392 (2002).
82. Egeblad, M. and Werb, Z. New functions for the matrix metalloproteinases in cancer progression. Nature Rev. Cancer 2, 161-174 (2002).

83. van de Schaft, D.W.J. *et al.* Effects of angiogenesis inhibitors on vascular network formation by human endothelial and melanoma cells. J. Natl. Caner Inst. 96(19):1473-1477 (2004).

Chapter 6

# DESIGN AND CONSTRUCTION OF ARTIFICIAL BLOOD VESSELS

Domenico Ribatti[1], Beatrice Nico[1], Elisabetta Weber[2]
*Department of Human Anatomy and Histology, University of Bari Medical School[1], Department of Neurosciences, Section of Molecular Medicine, University of Siena Medical School[2]*

## 1. ENDOTHELIUM AND MATRIX INTERACTIONS

Extracellular matrix (ECM) is a fibrillar meshwork of proteins, proteoglycans, and glycosaminoglycans. It constitutes a barrier for diffusion and convection, and provides tissue with mechanical strength and elastic properties. Turnover of the ECM is a unique biological problem because of the high collagen content of most ECM structures and the resistance of these triple helical molecules to the action of most proteases.

In arteries (Figure 6-1) and veins, the ECM constitutes more than half of the mass of the wall and is mainly composed of collagens and elastin (Figure 6-2). Quantitative biochemical studies of the collagen content of arterial tissue showed a high proportion of collagen type I and a lower amount of type III. Other components, such as fibronectin, microfibrils, proteoglycans and glycoproteins are all present within the extracellular spaces of the vessel wall and are crucial to vessel wall function and integrity. The extracellular material of the media, including collagen and elastic fibers, is produced primarily by the smooth muscle cells (SMC) during development. In the adventitia, collagen is synthesized and secreted by the adventital fibroblasts. As blood vessels develop during postnatal life, the vessel wall increases in diameter and thickness, accompained by a parallel increase in ECM and both elastin and collagen contents, as well as organization of the tunicae.



*R. Forough (ed.), New Frontiers in Angiogenesis,* 111–124.

Individual elastic fibers anastomose with each other forming net-like structures which spread predominantly in a circumferential direction. A more extensive degree of fusion occurs amongst SMC-producing lamellae of muscle cells, thus forming lamellar units, which allow the vessel to recoil after distension. An outer elastic lamina similar in appearance but markedly less well developed than the inner elastic lamina is situated at the outer aspect of the media at the boundary with the adventitia. Elastic fibers are less abundant in the surrounding adventitia. Collagen fibrils are found in all three tunicae and especially around the SMC of the media. Collagen is also abundant in the adventitia where it forms large bundles of fibrils which increase in size from the innermost region near the media to the outer most component.

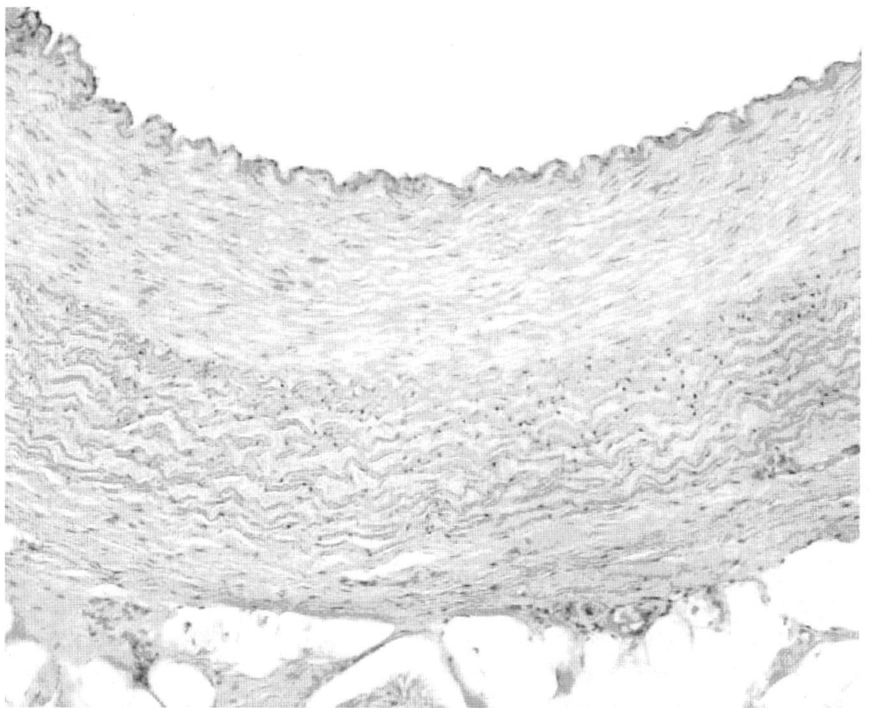

*Figure 6-1.* Hematoxylin-eosin staining of a canine aorta. Original magnification10x.

The activation of collagen is also regulated by cytokines and growth factors such as transforming growth factor beta (TGF-β), connective tissue growth factor, platelet derived growth factor (PDGF), epidermal growth factor (EGF) or fibroblast growth factor (FGF) produced by different cell types present in inflammatory sites. Cytokines, such as interleukin-1 (IL-1),

IL-4 and tumor necrosis factor alpha (TNF-$\alpha$), are also capable of modulating fibroblast proliferation and collagen deposition. Finally, hormones, such as aldosterone and angiotensin II stimulate collagen deposition.

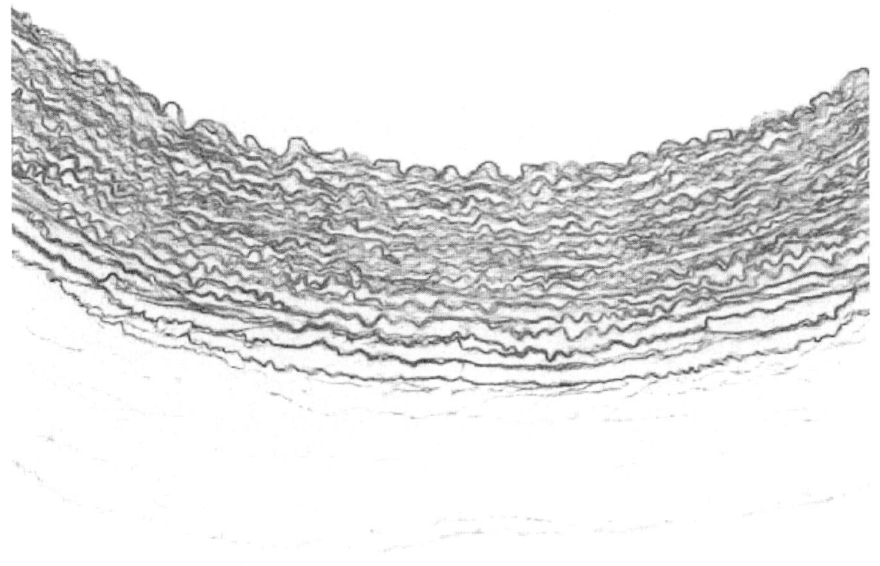

*Figure 6-2.* Weigert's staining of the elastic network of a rabbit aorta. Original magnification 10x.

The degradation of matrix involves a host of enzymes known as matrix metalloproteinases (MMP) which include collagenases, gelatinases, stromelysin, metalloelastases and membrane type-matrix metalloproteinases (MT-MMP), all of which can degrade specific matrix components.[1] These enzymes are produced by the vascular cell components including endothelial cells (EC) (Figure 6-3) SMC and fibroblasts, and are stored in a latent form within the ECM until they are required. They are capable of completely degrading the matrix, hence tightly-controlled mechanisms must be in place to regulate their activities.

The control of growth factor activity by ECM can be divided into distinct mechanisms: the release of growth factors from the matrix, and activation. Release involves the dissociation of growth factor from its matrix binding sites that results in the formation of a soluble pool of the growth factor. Heparan sulfate-bound growth factors, including PDGF, FGF-2 and vascular endothelial growth factor (VEGF), can be released from the matrix by heparin degrading enzymes. Activation refers to the dissociation of the mature growth factor from a latency protein or to induction of conformational changes in the growth factor that allows its binding to the receptors.

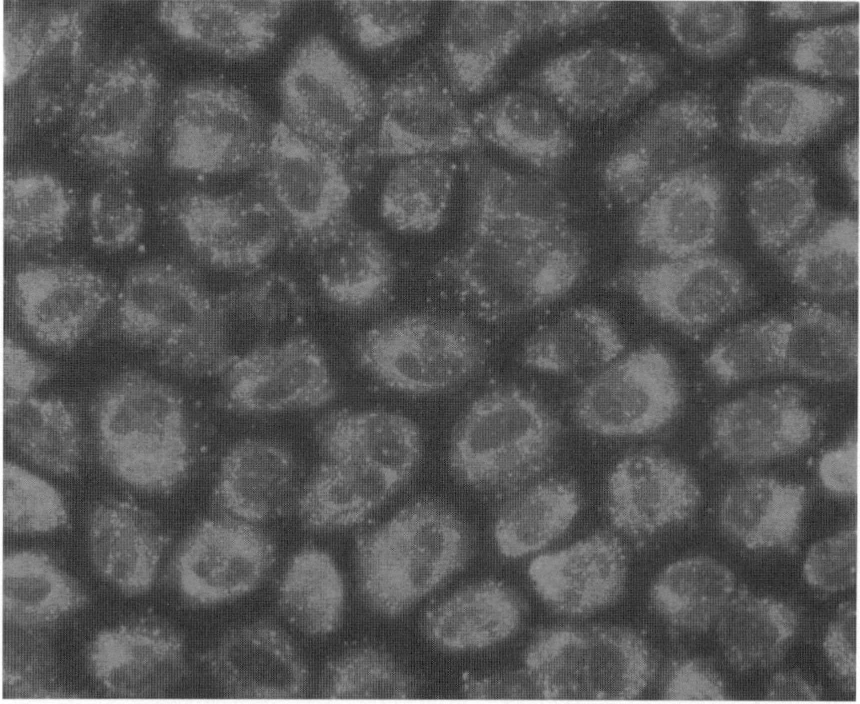

*Figure 6-3.* Staining of bovine aortic endothelial cells with DiI Ac-LDL. All cells stain brightly and uniformly for this endothelial marker. Original magnification 40x.

## 2. VASCULAR GRAFTS

Bypass surgery is a common treatment for coronary and peripheral vascular diseases. Autogenous vessels, particularly internal mammary arteries and saphenous veins of the leg, remain the standard for coronary grafting and peripheral bypass surgery.[2] However, up to 39% of patients do not have veins suitable for vascular reconstruction due to venous abnormality, poor quality and lack of vein due to previous surgery. As a result, alternative conduits constructed with synthetic materials have been investigated.

The use of synthetic materials to fabricate a blood vessel substitute began over three decades ago and has led to vascular prostheses made from a variety of materials. Biomaterials may be divided in two groups according to the time they require to function: the first group comprises biostable materials that should persist during the remaining life time of the patient, and the second group includes biodegradable biomaterials designed for use during a limited time period. Among all the types of biomaterials available, those of the polymer type have been used the most in clinical practice in the last years.

Polyethylene terephthalate (PET, Dacron) and expanded poly-tetrafluoroethylene (ePTFE) are currently the standard biomaterials of prosthetic vascular grafts in large diameter, high-flow, low-resistance locations, such as the aorta and the iliac and proximal femoral arteries, as coronary and peripheral bypass grafts in surgery and as A-V fistulae.

Both Dacron and ePTFE grafts have been shown to perform well at diameters $> 6$ mm, but neither material has been suitable for small-diameter ($< 4$ mm) applications, because they are compromised by thrombogenicity, infection, compliance mismatch the rigid graft and elastic host artery, fibrointimal hyperplasia, structural degeneration and aneurysmal dilation. When small vascular prostheses are grafted in areas of low blood flow, the prosthetic surface that makes contact with blood needs to be adapted by changing the structure of the prosthesis or coating its luminal surface with antithrombogenic agents or a cell layer as similar as possible to the natural blood vessel lining.

Finding a solution for small-diameter bypass grafting has become a major focus of attention. Tissue-based grafts are now considered to have potential for small diameter grafting applications. Several modifications to the basic graft have been proposed for improving its function. One is to increase the graft permeability on the basis of the notion that the rate of tissue ingrowth is associated with graft porosity (in limited porosity ranges) and that transmural capillary ingrowth can provide the source for the surface endothelialization.

## 3. TISSUE-ENGINEERED VASCULAR GRAFTS

Tissue-engineered vascular grafts may serve as an alternative source of vascular grafts. Tissue-engineering techniques try to improve different types of conduits by colonization with cells to achieve a live prosthesis with the capacity to take part in the tissue repair process.

Attempts have been made to decrease the surface thrombogenity of vascular grafts incorporating EC extracted from numerous sources onto the inner surface of prosthetic grafts. Engineered vessels must contain a confluent, adherent endothelial layer to be non thrombogenic, resistant to inflammation and neointimal proliferation. The ideal graft functions metabolically and biochemically like a native vessel, presenting a luminal substrate that optimizes healing and circulating cell adhesion. Although mechanical strength is a property that can be reproduced with passive materials, metabolic function requires cellular machinery.

Initial attempts at tissue engineering a blood vessel substitute involved seeding the lumen of a synthetic graft with endothelial cells. Seeding involves extracting autologous EC and then lining these cells onto the graft lumen. Sources of autologous tissues are as follows: i) veins; ii) arteries; iii) omental fat and iv) subcutaneous fat. Recent studies have assessed the possibility of using cells from the placenta, myofibroblasts from the umbilical cord or medullar stromal cells. All these cells show a strong capacity to differentiate into numerous mesenchymal cell lines, especially those of myofibroblastoid appearance. The general principles of extraction from these tissues have involved two main strategies: i) enzymatic loosening of EC from tissues and ii) separation of EC cells from non EC.

When prosthetic grafts seeded with EC are subjected to the action of pulsatile flow, poorly adhered cells become detached from the biomaterial surface, resulting in a 80-90% loss of these cells.[3] The shear stress provoked by the biomaterials is a significant factor in the host response. It is modulated by flow and affects protein absorption, leukocyte adhesion and thrombosis/coagulation.[4] The conformational changes that occur after platelet activation can have the consequence of exposing a larger surface area of receptors to a series of aggregation factors, thus increasing the thrombogenicity of the vascular prosthesis.[5] Contact of blood with the biomaterials surface can trigger the complement cascade and promote leukocyte adhesion. Leukocytes recognize proteins adsorbed by the biomaterial and may adhere to its surface through ligand-receptor adhesion and may release cytokine, growth factors and other bioactive agents that modulate the function of other cell types.[3]

Herring[6] was the first to report the successful isolation of EC and their subsequent transplantation on synthetic vascular grafts. Suspensions of isolated canine venous EC were used to precoat 6 mm Dacron graft before implantation in a canine model.[7] Four weeks after implantation, they observed 76% patency rates for seeded grafts compared to 22% rates for unseeded grafts.[8] Explanted grafts possessed an intact EC lining supported by SMC along with penetrating *vasa vasorum.*

EC adherence has been improved by conditioning vascular grafts to shear stresses before implantation[9] and by coating vascular graft lumens with EC-adherent proteins or sequences.[10-12] These peptides are based on the receptor-binding domains of cell adhesion proteins. Integrins, synthesized and expressed on the surface of EC, recognize the Arg-Gly-Asp (RGD) sequence.[13]

Holland et al.[14] showed increased *in vitro* EC adhesion, spreading and growth for cells plated on RGD attached to a starch-coated polystyrene surface compared with a fibronectin-coated surface.

Hubbell et al.[15] and Massia and Hubell[16] immobilized a synthetic peptide containing the sequence Arg-Glu-Asp-Val (REDV) on the otherwise non-adhesive glycophase glass and polyethylene terephatalate that had been surface-modified with polyethylene glycol. When the REDV sequence was immobilized on nonadhesive substrates, EC attached and spread, but fibroblasts, SMC and platelets did not. They also found that the endothelial monolayers on REDV-grafted substrates were nonthrombogenic.

Tiwari et al.[17] found that RGD and heparin that had covalently bounded to Myolink (Credent Vascular Technologies, Ltd, Wrexham, Koles, UK) graft surfaces improved cell reduction and provided an anti-thrombogenic surface for initial blood flow *in vivo.*

Another method for creating blood vessels involves the co-culture of cells with natural and/or synthetic materials in order to create cellular vessels *in vitro.*[18] The first complete tissue-engineered biological conduit was constructed in 1986 by Weinberg and Bell[19], who developed a vascular conduit composed of SMC cultured in a collagen tube. A layer of fibroblasts was then added around the outside of the tube, and EC were seeded on the luminal surface. A Dacron sleeve was added between the medial and adventitial layers to provide strength to withstand physiological pressure. Electron microscopy showed that EC lining the lumen and the SMC in the wall were healthy, well differentiated and biochemically active, producing von Willebrand factor (vWF) and prostacyclin after 1 week in culture. Although the graft obtained 92% EC coverage on the inner surface and longitudinal SMC organization, it failed to show the requisite mechanical strength.

Pediatric patients with cardiovascular malformations often require substitution of pulmonary vessel segments and, since prosthetic materials lack growth potential, replacement of grafts is required as the children grow. Pulmonary circulation, due to its low pressure, is an ideal setting to test tissue engineered blood vessels. A tissue engineered pulmonary artery conduit was created by seeding tubular polyglactin/polyglycolic acid (PGA) scaffolds with SMC followed by EC.[20] After 7 days of an *in vitro* culture, the conduits were implanted into the ovine pulmonary arteries and evaluated between 11 and 24 weeks. In contrast to the acellular control, the tissue engineered scaffolds appeared to be histologically nearly identical to native arteries.

For the first time in 2001, a biodegradable polymer scaffold (poly-caprolactone-polylactic acid copolymer reinforced with woven polyglycolic acid), seeded with autologous endothelial cells collected from a peripheral vein and expanded *in vitro*, was implanted in a four year old girl. Seven months after implantation there was no evidence of graft occlusion or aneurysm formation.[21] The same technique was later modified seeding the grafts with autologous aspirated bone marrow cells with the advantages of avoiding extra-hospitalization for vein harvesting and the risk of infection associated with cell culturing.[22] The 22 patients of the study had no thrombogenic complications, stenosis or obstruction of tissue-engineered autografts. When systemic circulation arterial segments are to be substituted, the high pressure is undoubtedly a great challenge for tissue engineered blood vessels. The use of bioreactors to simulate blood flow conditions seems of great value in strengthening the vessel wall and orienting smooth muscle cells and collagen fibers in circular arrays as in native arteries.

Niklason et al.[23] grew blood vessels in polymeric tubes under flow conditions. Bovine aortic SMC were plated on the outside of tube-shaped PGA scaffolding and placed in a silicone tube circuit in the presence of pulsatile flow, i.e. similar to flow in native blood vessels. After 8 weeks, the grafts were removed and the surfaces were coated with bovine aortic EC. Pulsed grafts were thicker, had greater suture retention, more physiological SMC and collagen density and remained patent longer. Similar findings were observed when porcine saphenous arterial xenografts were compared to non-pulsed engineered grafts. The vessel structure displayed a contractile response to vasoconstrictors. These tissue engineered arteries, implanted in the saphenous artery (a branch of the femoral artery) of miniature swine, remained patent for at least one month.

The importance of applying shear stress conditions to tissue-engineered blood vessels designed for systemic circulation has been highlighted by several studies. A recent work by Opitz et al.[24] compared grafts incubated either in a pulsatile flow bioreactor or under static conditions. The grafts were obtained by seeding enzymatically derived vascular SMC onto bioabsorbable P-4-HB (poly-4-hydroxybutyrate) scaffolds. After a 14-day incubation in dynamic or static conditions, they were luminally seeded with EC. Bioreactor-cultured vessels showed complete cellular coverage of the scaffold, had more protein content, higher cell numbers and enhanced ECM protein synthesis, compared to static controls. Even if bioreactor culturing increases the strength of tissue engineered constructs, native vessels are significantly more elastic. Following degradation of the polymeric scaffold, tissue-engineered blood vessels implanted in the aorta of sheep, due to insufficient elastic fiber synthesis, were unable to withstand blood pressure and underwent dilatation.[25]

Campbell et al.[26] inserted silastic tubing of variable dimensions in the peritoneal cavity of rats or rabbits. In that position, the inflammatory reaction covered the tubes with successive layers of myofibroblasts, collagen matrix, and a monolayer of mesothelial cells. These structures could be removed from the silastic tubes and converted to create a synthetic artery whose architecture mirrored that of a normal vessel, with the mesothelial cells serving as the endothelium, the myofibroblasts as SMC analogues in a bed of collagen and elastin, as well as collagenous adventitia. These structures were interpostionally grafted end-to-end into rabbit carotid arteries or rat abdominal aortas in the very same animals in which they had grown and remained patent for at least four months. These authors demonstrated that it is possible to remove the immune reaction that may follow cell-based grafting by culturing tissue in the anticipated host.

Promising scaffolds come from the field of bionanotechnology.[27] It is known that cells attach and organize well around fibers with diameters smaller than their diameter. Aligned biodegradable nanofibrous structures of copolymer [P(LLA.CL)], made by the technique of electrospinning, promoted adhesion, proliferation and alignment of SMC with a final topography resembling the circumferential orientation of cells and fibers in the native arterial media.

Several techniques have been proposed to improve cell adhesion and proliferation on the scaffolds. Fibrin matrices have been conjugated with VEGF to attract EC, stimulate EC proliferation and enhance EC attachment.[28] In addition, VEGF-producing SMC has found to promote EC proliferation and migration.

## 4. COMPLETELY BIOLOGICAL TISSUE-ENGINEERED BLOOD VESSELS

In another system, EC covered 99.2% of the luminal surface and displayed differentiated properties *in vitro*, including expression of von vWF, of functional thrombin receptors, synthesis of prostacyclin and inhibition of platelet aggregation *in vitro*.[29] These authors reported a novel approach that was based exclusively on the use of cultured human cells without any synthetic or exogenous biomaterials. This paper is the first report of a completely biological tissue-engineered blood vessels to display a burst strength comparable to that of human vessels (> 200 mmHg). SMC and fibroblasts were cultured for 5 weeks in the presence of vitamin C to create a three-dimensional ECM with characteristics similar to that observed *in vivo*. It has been shown that SMC grown in cultures supplemented with ascorbate synthesized three times as much collagen as SMC incubate without ascorbate.[30] This cohesive cellular sheet was placed around a tubular support to produce the media of the vessel. A similar sheet of fibroblasts was wrapped around the media to produce the adventitia. The construct was plated for a maturation period of 56 days in a bioreactor, designed to provide perfusion of the culture medium and mechanical support. The mechanical stimuli for pulsatile flow could generate the cyclic strain necessary to alter ECM production, thereby creating a histologically organized, functional construct with satisfactory mechanical characteristic for implantation. Thereafter, an integrated tubular structure could be removed. This technique resulted in a well-defined, three layered organization and production of ECM proteins, including elastin. Moreover, SMC re-expressed desmin. This structure was implanted as a canine femoral arterial interposition graft and remained patent in 3 of 6 animals at one week.

## 5. DECELLULARIZED BIOLOGICAL SCAFFOLDS

The preparation of natural decellularized grafts is a current topic of research. The cells with their surface antigens are removed by means of detergent and enzymatic extraction methods, leaving a well-preserved acellular matrix that provides a scaffold for autologous cell ingrowth and allows favorable tissue remodeling. Treatment with detergents removes proteoglycans and, in turn, reduces the binding of the latter to collagen, generating a porous matrix that promotes the deposition of calcium salts.[31] In addition, the presence of cell debris, lipid and soluble proteins contribute towards the appearance of a host immune response after graft implant.[32] These undesirable effects could be resolved by stabilizing the biomaterial through cross-linking techniques.

Allogenic scaffolds have achieved minimal immunoreactivity and good durability for as long as 6 years, without aneurysm degradation.[33] Xenogenic acellular scaffolds elicit significant chronic immunoresponsive inflammation, which is sufficient to destroy elastin structures.[32,34] Scaffolds constructed from decellularized porcine intestine with cross-linked type I collagen deposited on the luminal surface have also been tested. The constructs provided the necessary mechanical and hemodynamic properties at inflammation. The scaffolds were cellularized and EC were oriented in the direction of blood flow within three months in rabbit carotid artery bypass graft models.[35]

A decellularized biologic extracellular matrix produced by human fibroblasts in culture was recently designed for replacement of small-diameter vessels.[36] The interior of the vessel was covered by endothelial cells. The resulting vessel had an endothelium which had good hemostatic parameters (vWF secretion and thrombomodulin expression) and low levels of adhesive molecules (E-selectin and ICAM-1), that usually correlate with rejection in organ transplantation.

## 6. MECHANICAL STRENGTH

The mechanical strength of a vessel wall is derived largely from ECM components, particularly collagen and elastin, which are produced by SMC in the medial layer of the artery wall. Collagen, which is the major factor responsible for mechanical strength of engineered vessels, is present throughout the arterial wall. Its native state is resistant to most proteases but is readily degraded by a wide variety of proteases once denatured. Increasing intramolecular cross-links among its three consisting peptide subunits can increase the tensile strength of collagen fibers and make it less susceptible to degradation. Collagen facilitates cell attachment with its integrin-binding domains and the collagen matrix supports tissue growth.

The addition of cyclic strains[37], growth factors[38], and ascorbic acid[39] to cells in culture have been shown to increase collagen transcription *in vitro*. The formation of cross-links between and within collagen fibrils stabilized the proteins, making them less susceptible to attack by matrix metalloproteinases and other enzymes.[40]

## ACKNOWLEDGEMENTS

Supported in part by Fondazione Italiana per la Lotta al Neuroblastoma, Genoa, Ministero della Salute – Ricerca Finalizzata FSN 2002, Rome and Ministero dell'Istruzione, dell'Università e della Ricerca (Consorzio Carso No 72/2), Rome, Italy.

## REFERENCES

1. Visse R, Nagase H. Matrix metalloproteinases and tissue inhibitors of metalloproteinases: structure, function, and biochemistry. Circ Res, 92: 827-839, 2003.

2. Eagle K, Guyton RA, Davidoff R, Ewy GA, Fonger J, Gardner TJ, Gott JP, Hermann HC, Marlow RA, Nugent WC, O'Connor GT, Orszulak TA, Rieselbach RE, Winters WL, Yusuf S. ACC/AHA guidelines for coronary artery bypass graft surgery. J Am Coll Cardiol , 34: 1262-1347, 1999.
3. Rosenman JE, Kempczinski RF, Pearce WH, Silberstein EB. Kinetics of endothelial cell seeding. J Vasc Surg, 2: 778-784, 1985.
4. Kao WJ. Evaluation of leukocyte adhesion on polyurethanes: the effects of shear stress and blood proteins. Biomaterials, 21: 2295-2303, 2000.
5. Mikuki SA, Greisler H. Understanding and manipulating the biological response to vascular implants. Semin Vasc Surg, 12: 18-26, 1999.
6. Herring M. A single-staged technique for seeding vascular grafts with autogenous endothelium. Surgery, 84: 498-504, 1978.
7. Herring M, Gardner A, Glover J. Seeding endothelium onto canine arterial prostheses. The effects of graft design. Arch Surgery, 114: 679-682, 1979.
8. Herring MB, Dilley R, Jersild RA Jr, Boxer L, Gardner A, Glover J. Seeding arterial prostheses with vascular endothelium. The nature of the lining. Ann Surg, 190: 84-90, 1979.
9. Ott M, Ballermann BJ. Shear stress-conditioned, endothelial cell-seeded vascular grafts improved cell adhesion adherence in response to in vitro shear stress. Surgery, 117: 334-339, 1995.
10. Foxall TL, Auger KR, Callow AD, Libby P. Adult human endothelial cell coverage of small-caliber Dacron and polytetrafluoroethylene vascular prostheses in vitro. J Surg Res, 41: 158-172, 1986.
11. Zilla P, Fasol R, Preiss P, Kadletz M, Deutsch M, Schima H, Tsangaris S, Groscurth P. Use of fibrin glue as a substrate of PTFE vascular grafts. Surgery, 105: 512-522, 1989.
12. Thomson GJ, Vohra RK, Carr MH, Walker MG, 1991. Adult human endothelial cell coverage of small-caliber dacron and polytetrafluoroethylene vascular grafts: a comparison of four substrates. Surgery, 109: 20-27, 1991.
13. Cheresh DA. Human endothelial cells synthesize and express an Arg-Gly-Asp directed adhesion receptor involved in attachment to fibrinogen and von Willebrand factor. Proc Natl Acad Sci USA, 94: 6471-6475, 1987.
14. Holland J, Hersh L, Bryham M, Onyiriuka E, Ziegler L. Culture of human vascular endothelial cells on a RGD-containing synthetic peptide attached to a starch-coated polystyrene surface: comparison with fibronectin-coated tissue grade polystyrene. Biomaterials, 17: 2147-2156, 1996.
15. Hubbell JA, Massia SP, Desai NP, Drumheller PD. Endothelial cell-selective materials for tissue engineering in the vascular graft via a new receptor. Biotechniques, 9: 568-573, 1991.
16. Massia SP, Hubbell JA. Vascular endothelial cell adhesion and spreading promoted by the peptide REDV of the IIICS region of plasma fibronectin is mediated by integrin $\alpha 4\beta 1$. J Biol Chem, 267: 14019-14026, 1992.
17. Tiwari A, Salacinski HJ, Punshon G, Hamilton G, Seifalian AM. Development of a hybrid cardiovascular graft using a tissue engineering approach. FASEB J, 16: 791-796, 2002.
18. Edelman ER. Vascular tissue engineering: designer arteries. Circ Res, 85: 1115-1117, 1999.
19. Weinberg CB, Bell E. A blood vessel model constructed from collagen and cultured vascular cells. Science, 231: 397-400, 1986.

20. Shinoka AT, Shum-Tim D, Ma PX, Tanel RE, Isogai N, Langer R, Vacant JP, Mayer JE Jr. Creation of viable pulmonary artery autografts through tissue engineering. J Thorac Cardiovasc Surg, 115: 536-545, 1998.
21. Shin'oka AT, Imai Y, Ikada Y. Transplantation of a tissue-engineered pulmonary artery. New Engl J Med, 344:532-3, 2001.
22. Matsumura G, Hibino N, Ikada Y, Kurosawa H, Shin'oka T. Successful application of tissue engineered vascular autografts: clinical experience. Biomaterials 24:2303-2308, 2003.
23. Niklason LE, Gao J, Abbott WM, Hirschi KK, Houser S, Marini R, Langer R. Functional arteries growth in vitro. Science 284: 489-493, 1999.
24. Opitz F, Schenke-Layland K, Ritcher W, Martin DP, Degenkolbe I, Wahlers T, Stock UA. Tissue engineering of bovine aortic blood vessel substitutes using applied shear stress and enzymatically derived vascular smooth muscle cells. Annals of Biomedical Engineering, 32:212-22, 2004.
25. Opitz F, Schenke-Layland K, Cohnert TU, Starcher B, Halbhuber KJ, Martin DP, Stock UA. Tissue engineering of aortic tissue: direct consequence of suboptimal elastic fiber synthesis in vivo. Cardiovasc. Res. 63:719-730, 2004.
26. Campbell JH, Efendy JL, Campbell GR. Novel vascular graft grown within recipient's own peritoneal cavity. Circ Res, 85: 1173-1178, 1999.
27. Xu CY, Inai R, Kotaki M, Ramakrishna S. Aligned biodegradable nanofibrous structure: a potential scaffold for blood vessel enginnering. Biomaterials 25:877-886, 2004.
28. Zisch AH, Schenk U, Schense JC, Sakiyama-Elbert SE, Hubbell JA. Covalently conjugated VEGF-fibrin matrices for endothelialization. J Controlled Release, 72: 101-113, 2001.
29. L'Heureux N, Paquet S, Labbé R, Germain L, Auger FA. A completely biological tissue-engineered human blood vessel. FASEB J, 12: 47-56, 1998.
30. Barone LM, Faris B, Chipman SD, Toselli P, Oakes BW, Franzblau C. Alteration of the extracellular matrix of smooth muscle cells by ascorbate treatment. Biochim Biophys Acta, 840: 245-254, 1985.
31. Jackson RL, Busch SJ, Cardin AD. Glycosaminoglycans: molecular properties, protein interactions and role in physiological processes. Physiol Rev, 71: 481-539, 1991.
32. Courtmann DW, Errett BF, Wilson GJ. The role of crosslinking in modifications of the immune response elicited against xenogenic vascular acellular matrices. J Biomed Mater Res, 55: 576-586, 2001.
33. Wilson FJ, Yeger H, Klement P, Lee JM, Courtmant DW. Acellular matrix allograft small caliber vascular prostheses. ASAIO Trans, 36: M340-M343, 1990.
34. Bader A, Steinhoff G, Strobl K, Schilling T, Brandes G, Mertsching H, Tsikas D, Froelich J, Havenich A. Engineering of human vascular aortic tissue based on a xenogenic starter matrix. Transplantation, 70: 7-14, 2000.
35. Huynh T, Abraham G, Murray Y, Brockbank K, Hagen PO, Sullivan S. Remodeling of an acellular collagen graft into a physiologically responsive neovessels. Nature Biotechnol, 17: 1083-1086, 1999.
36. Remy-Zolghadri M, Laganiere J, Oligny JF, Germain L, Auger FA. Endothelium properties of a tissue-engineered blood vessel for small-diameter vascular reconstruction. J.Vasc.Surg. 39: 613-620, 2004.
37. Leung DYM, Glagow S, Mathews MB. Cyclic stretching stimulates synthesis of matrix components by arterial smooth muscle cells in vitro. Science, 191: 475-477, 1976.
38. Varga J, Rosenbloom J, Jimenez SA. Transforming growth factor β (TGFβ) causes a persistent increase in steady-state amounts of type I and type III collagen and

fibronectin mRNAs in normal human dermal fibroblasts. Bicohem J, 247: 597-604, 1987.
39. Geesin JC, Darr D, Kaufman R, Murad S, Pinnel SR. Ascorbic acid specifically increases type I and type III procollagen messanger RNA levels in human skin fibroblasts. J Invest Dermatol, 90: 420-424, 1988.
40. Eyre D. Cross-linking in collagen and elastin. Annu Rev Biochem 34: 1262-1347, 1984.

Chapter 7

# BIOENGINEERING ANGIOGENESIS: NOVEL APPROACHES TO STIMULATING MICROVESSEL GROWTH AND REMODELING

Richard J. Price, Meghan M. Nickerson, John C. Chappell, Christoper R. Anderson, Ji Song
*Department of Biomedical Engineering, University of Virginia*

## 1. INTRODUCTION

Angiogenesis, the growth of new blood vessels from existing blood vessels, and arteriogenesis, the formation and expansion of arterioles and arteries, are closely related processes that directly impact countless physiological and pathological conditions. Because of their ubiquitous nature, angiogenesis and arteriogenesis offer a number of unique challenges for the future of health care, with perhaps the most visible challenge stemming from the fact that tumors utilize angiogenesis to facilitate both the growth and metastasis of cancer. However, angiogenesis and arteriogenesis also offer exciting therapeutic opportunities, and the development of novel methods for mobilizing our natural capacity to grow and remodel the microvasculature will ultimately have a tremendous impact on medicine in the future. Neovascularization therapy will have a significant bearing on the success of tissue engineering and on our ability to restore blood flow to tissues and organs affected by occlusive vascular disease, particularly for those patients that are not amenable to surgical revascularization.

Although neovascularization therapy retains great promise for these applications, it is also clear that new insights will be required to bring these approaches to broad clinical usage. To date, in contrast to numerous animal studies, clinical trials for angiogenesis therapy based on growth factor protein and gene delivery have yielded mixed results at best.[1,2] These



*R. Forough (ed.), New Frontiers in Angiogenesis,* 125–157.

disappointing results are due, at least in part, to the fact that the growth factors are typically not delivered in a controlled and/or sustained fashion. More recent trials based on stem and progenitor cell delivery have shown promise, but they are still in their infancy.[3] With regard to tissue engineering, the need for a robust vascular supply for many engineered tissues, especially those with high metabolic requirements, is now well recognized in the field.[4] Overcoming these challenges will likely require novel combinations of techniques and approaches from many disciplines, including chemistry, genetic engineering, molecular biology, chemical engineering, and mechanical engineering. Given the potential need for such multi-disciplinary strategies, it is clear that bioengineers are uniquely poised to make substantial contributions to the field. In this chapter, we will explore cutting-edge bioengineering applications for angiogenesis. The chapter essentially consists of two major sections. The first presents bioengineering approaches to therapeutic neovascularization, while the second is focused on strategies for eliciting vessel growth in tissue engineering applications.

## 2. BIOENGINEERING THERAPEUTIC ANGIOGENESIS AND ARTERIOGENESIS

Many new bioengineering technologies for stimulating therapeutic neovascularization through the delivery of pro-angiogenic growth factors and cells and aimed at overcoming many of the limitations of current strategies, are beginning to emerge. For this chapter, they have been divided based on means of delivery, namely implantation or injection. Implantable delivery systems, consisting of controlled release systems, which typically involve the use of biodegradable polymer to release growth factor, and tissue engineered constructs, which consist of cells grown on synthetic polymers or native extracellular matrix, are discussed in sections *2.1* and *2.2*, respectively. Table 7-1 then summarizes the studies that are presented in *2.1* and *2.2*. Injectable delivery systems are then described in sections *2.3* and *2.4.*, with section *2.3* devoted to systemic vascular injection strategies and section *2.4* devoted to intramuscular and intramyocardial injection strategies.

*Table 7-1.* Implantable Controlled Release Systems for Stimulating Neovascularization.

| Polymer/Construct Carrier | Growth Factor(s) Delivered | Target Tissue, Organ, or Cell | Reference |
|---|---|---|---|
| gelatin hydrogel | aFGF | rat subcutis | Thompson et al. (1988)[6] |
| gelatin hydrogel | bFGF | mouse subcutis | Tabata et al. (1994)[7] |
| gelatin hydrogel | bFGF | mouse subcutis | Tabata and Ikada (1999)[6] |
| alginate hydrogel | VEGF | mouse ischemic hindlimb | Lee et al. (2000)[8] |
| alginate hydrogel | VEGF/bFGF | mouse subcutis | Lee et al. (2003)[10] |
| PLG | VEGF | endothelial cells in vitro | Murphy et al. (2000)[11] |
| PLG | VEGF | HMVEDs/ mouse subcutis | Peters et al. (2002)[12] |
| PLG | VEGF | hepatocytes/mouse subcutis | Smith et al. (2004)[13] |
| EVA | VEGF/PDGF | mouse ischemic hindlimb | Richardson et al. (2002)[14] |
| EVA | multiple | mouse subcutis | Fajardo et al. (1998)[15] |
| EVA | aFGF | infarcted procine myocardium | Lotez et al. (1998)[17] |
| amylopectin | bFGF | mouse subcutis | Tabata et al. (1998)[18] |
| collagen hydrogel | VEGF | mouse subcutis | Tabata et al. (2000)[19] |
| fibrin | VEGF | claudicated human limbs | Kipshidze et al. (1999)[20] |
| 3DFC | multiple | endothelial cells in vitro | Pinney et al. (2000)[12] |
| 3DFC | multiple | infarcted mouse myocardium | Kellar et al. (2001)[23] |
| BAM | VEGF/multiple | ischemic mouse hindlimb | Lu et al. (2001)[24] |
| BAM | VEGF/multiple | sheep myocardium | Lu et al. (2002)[25] |

aFGF=acidic fibroblast growth factor, bFGF=basic fibroblast growth factor, VEGF=vascular endothelial growth factor, PLG=poly(lactide-co-glycolide), HMVEC=human microvascular endothelial cells, PDGF=platelet derived factor, 3DFC-three-dimensional fibroblast culture, BAM=bioartificial muscle.

## 2.1 Implantable Therapeutics: Controlled Release Systems

As noted above, clinical neovascularization trials centered on growth factor protein and gene delivery have yielded only mixed results that, for the most part, have not lived up to the promise and potential generated by animal studies. One critical factor in the apparent lack of success of these trials is that the dosage of the applied growth factor may be too transient to elicit a measurable neovascularization response. Indeed, angiogenesis and arteriogenesis are complicated processes that may require several weeks or even months to occur and withdrawal of a stimulus during this time could be detrimental. In response to the need for sustained delivery of growth factors in neovascularization therapy, several controlled release strategies which invoke the use of biocompatible polymers are under development. Of these, most strategies are based on the use of one of three basic polymers, namely gelatin hydrogels, alginate, and poly(lactide-co-glycolide). While the remainder of this section will focus on controlled release systems based on these three polymers, other formulations, including collagen hydrogels, amylopectin, and polyvinyl chloride, will also be discussed.

### 2.1.1 Gelatin Hydrogels

To date, one of the most widely used delivery systems for angiogenic growth factors has been hydrogels containing the naturally derived polymer gelatin. Gelatin hydrogels are an advantageous delivery polymer for several reasons. Polymer hydrogel is relatively inert with respect to the protein it carries, and it has been widely used in a many clinical applications, so safety issues are minimal. Moreover, polymer hydrogels permit fine tuning of biodegradation and growth factor release rates through modifications in gelatin content, gelatin concentration and, in the case of cross-linked polymers, the concentration of the cross-linker itself, such as glutaraldehyde.[5]

Using primarily members of the fibroblast growth factor family as a stimulus, the use of gelatin hydrogels for generated tissue neovascularization has been tested in several animal models and with multiple modifications of the gelatin hydrogel construct itself. Thompson et al.[6] described the delivery of acidic FGF in gelatin hydrogels and site-directed neovascularization in rats. Gelatin hydrogels have been used for the delivery of bFGF in several applications as well. Tabata et al.[7] created bFGF-impregnated gelatin hydrogel disks and implanted them into mice subcutaneously. In contrast to bolus bFGF injection and bFGF-free hydrogel, a significant increase in vascularity in tissue surrounding the implant, as assessed by tissue hemoglobin content, was observed. Moreover, this response was found to be dose dependent, with maximal vascularization occurring at 100 μg bFGF per mouse. In an important extension of these studies, this same group later tested the effect of gelatin hydrogel biodegradability on neovascularization.[5] Using acidic hydrogels with an isolelectric point of 5.0 and basic hydrogels with an isoelectric point of 9.0 as a base, the authors tested how complexation of bFGF in the acidic hydrogel differed with respect to vascularization in comparison to the basic hydrogel in which no complexation occurred and bFGF was released almost immediately. Their *in vivo* results indicated that biodegradation of the acidic hydrogel created a more significant and sustained vascularization response, demonstrating the ability to control surrounding vascular density through manipulations of the hydrogel properties. More recently, this group has incorporated hepatocyte growth factor (HGF) into their implantable gelatin hydrogel system. When implanted subcutaneously, HGF gelatin hydrogels elicited a significant increase in VGEF expression in the surrounding tissue, as well as an increase in capillarity when compared to bolus HGF injection and blank hydrogel.

### 2.1.2 Alginate Hydrogels

Another naturally derived polymer that has received considerable use for controlled release of angiogenic growth factors is alginate. Alginate is a linear polysaccharide copolymer of (1-4)-linked β-D-mannuronic acid and α-L-guluronic acid and is derived primarily from bacteria and brown seaweed.[8] Divalent cations interact with the α-L-guluronic acid monomers to form bridges, thereby transitioning the material into a gel. The mechanical properties, including strength and pore size, of alginate gels are controlled by altering the amount of each monomer. Alginate gels slowly dissolve *in vivo* as $Ca^{++}$ is lost, causing the dissociation of individual polymer chains. Because of their wide availability, lack of toxicity, and flexibility with regard to chemical and mechanical properties, alginate gels are attractive as controlled release systems for therapeutic angiogenesis.

While much of the work in the use of alginate hydrogels revolves around "injectable" systems because of small bead geometries. (Note: these strategies are appropriately described later in the Chapter under the heading *Injectable Therapeutics*), implantable forms have also enjoyed success. One particularly interesting application of alginate hydrogels for neo-vascularization employs the release of growth factor in response to mechanical stimulation.[9] After establishing that mechanical compression increases both the rate and amount of released VEGF in a compression amplitude dependent manner, VEGF loaded alginate hydrogel disks were implanted into ischemic hindlimbs of non-obese diabetic mice and blood vessel density was measured after 2 weeks. VEGF hydrogels elicited an increase in vessel density in comparison to control hydrogels without VEGF; however, even greater increases in vessel density were observed when VEGF hydrogels were exposed to mechanical compression. Later, this same group used alginate hydrogels to compare neovascularization in tissues surrounding alginate hydrogel disks impregnated with VEGF and bFGF.[10] Their results indicate that, while bFGF creates a thicker granulation tissue layer around the hydrogel and more total vessels, VEGF generates a greater density of vessels.

### 2.1.3 Poly(lactide-co-glycolide) Scaffolds

The synthetic polymers poly(lactic acid), poly(glycolic acid), and their copolymers have played a central role in numerous tissue engineering applications. These polymers are particularly useful for tissue engineering applications because they can be used for the sustained release of biologically active proteins, they degrade into natural metabolites, and they can be processed into different geometries to generate scaffolds for cell adhesion. Moreover, these manipulations may be done with exceptional reproducibility. Using poly(lactide-co-glycolide) (PLG) as a base polymer, Murphy et al.[11] developed a tissue engineering scaffold that slowly released VEGF during polymer degradation. To demonstrate biological activity of the released VEGF, human dermal microvascular endothelial cells were shown to undergo significantly increased cell proliferation. Peters et al.[12] and Smith et al.[13] then used a similar VEGF bearing PLG construct, in conjunction with transplanted HMVECs and hepatocytes, respectively, to demonstrate enhanced angiogenesis within the construct itself. These studies are described in more detail in Section 3 when angiogenesis in tissue engineering is presented. They are mentioned here because the VEGF/PLG polymer used in these studies was modified by the addition of a second factor (PDGF-BB) by Richardson et al.[14] and shown to elicit a significant therapeutic effect. Specifically, Richardson et al.[14] designed a construct that created a relatively rapid release of VEGF and a relatively slow release of PDGF-BB. The goal was to first stimulate rapid new vessel growth with VEGF and then elicit maturation of these new vessels with PDGF-BB, which primarily acts as a mitogen and chemoattractant for smooth muscle cells and pericytes. Interestingly, in addition to showing that this construct became invested with mature microvessels, the authors also demonstrated that, when the constructs were implanted into the ischemic hindlimbs of non-obese diabetic mice, tissue regions surrounding the construct exhibited significant arteriogenesis. Although changes in perfusion were not measured, these results do indicate the potential for using PLG based growth factor delivery systems for therapeutic arteriogenesis.

### 2.1.4 Other Polymers

A number of other synthetic and natural polymers have also been used to deliver angiogenic growth factors, typically in the context of model angiogenesis systems or with the goal of stimulating therapeutic vessel growth. Synthetic ethyl-vinyl acetate (EVA) copolymer has been used for many years as a controlled delivery material[15] and as a model system for studying angiogenesis.[16] Furthermore, of particular importance for this topic, EVA polymer has been used as a perivascular source of aFGF in a porcine model of chronic myocardial ischemia to improve myocardial flow and left ventricular function.[17] Amongst naturally occurring polymers, amylopectin hydrogel disks bearing bFGF have been shown to induce blood vessel growth subcutaneously[18], collagen hydrogels impregnated with VEGF stimulate new vessel growth[19], and fibrin glue has been used as a VEGF carrier in an attempt to ameliorate claudication in human patients.[20]

## 2.2 Implantable Therapeutics: Tissue Constructs

During physiological and pathological processes, such as those that occur in skeletal muscle during exercise, in the female reproductive system, and in response to wound healing, angiogenesis is coordinated by parenchymal, inflammatory, stem, and vascular cells. For this reason, the use of cells as stimulants for therapeutic neovascularization, particularly as natural sources of angiogenic growth factors, is a rational strategy. Indeed, as described in other chapters of this book, marrow-derived cells, endothelial derived precursor cells, and other stem cells are being employed for this purpose. These cells are typically delivered by direct injection to the intended site of action. However, it is probable that, in some scenarios, the effectiveness of these cells can be amplified by more specifically localizing them to selected regions, increasing their total numbers, and/or performing ex vivo genetic manipulations. To some extent, each of these qualities may be enhanced by delivering cells to sites of ischemia by adhering them to scaffolds and generating tissue engineered constructs. In a later section, the vascularization of tissue engineered constructs will be addressed. Here, the use of such constructs for therapeutic neovascularization of surrounding tissues is presented.

### 2.2.1 Three Dimensional Fibroblast Culture (3DFC)

The 3DFC tissue construct generated by Advanced Tissue Sciences essentially consists of fibroblasts grown on a three-dimensional matrix comprised of a lactate/glycollate synthetic copolymer. In early studies[21], subcutaneous implantation of 3DFCs suggested that they may have pro-angiogenic activity. In 2000, Pinney et al.[22] followed up on these early results and performed a number of experiments aimed at determining levels of secreted growth factor from 3DFCs. In comparison to cultured monolayers, 3DFC constructs secreted 3-5 fold more HGF and VEGF *in vitro* and conditioned medium from 3DFCs elicited significantly enhanced microvascular endothelial cell proliferation. When the angiogenic activity of 3DFCs was tested using a chick chorioallantoic membrane assay, a substantial increase in branch points, which was ameliorated by antibodies to VEGF and HGF, was seen with 3DFC in comparison to scaffold control. To then further extend these results to a mammalian model of myocardial infarction, Kellar et al.[23] implanted 3DFCs onto hearts of immunodeficient mice that had undergone coronary artery occlusion. 3DFCs with non-viable cells, untreated ischemic hearts, and normal hearts served as controls. After both 14 and 30 days, 3DFCs exhibited a more than 2-fold increase in microvessel density over the infarct controls, with the depth of neovascularization being approximately 1/3 of the ventricle wall thickness. Analysis with SM α-actin staining revealed that the percentage of vessels that were arterioles was substantially enhanced as well, supporting the ability of 3DFCs to stimulate a therapeutic arteriogenic response.

### 2.2.2 Bioengineered Skeletal Muscle Constructs

Another novel cell-based bioengineering approach to this problem utilizes bioengineered artificial muscle (BAM) constructs that have been genetically engineered to secrete high levels of growth factor. In contrast to other approaches where genetically modified myoblasts are injected and cell survival is an issue, BAMs survive for long periods of time and secrete reasonably predictable levels of growth factor.[24] The therapeutic potential of BAMs that have been engineered to secrete high levels of recombinant VEGF has been tested *in vivo* using both a hindlimb ischemia model for mice and sheep myocardium. In the mouse, following ligation of the femoral and saphenous arteries, VEGF secreting BAMs were placed over the tibialis muscle. While no measurements of perfusion restoration or arteriogenesis were made in this study, a 2-3 fold increase in capillary density was observed in the underlying tibialis muscle treated with an implanted BAM.[24] In sheep myocardium, VEGF secreting BAMs were implanted at multiple sites using several approaches, including fibrin glue, suturing, and direct insertion using an angiocatheter. Here, using immunochemistry, the authors located increased VEGF in myocardial regions near implanted BAMs, as well as a considerable increase in capillary density.[25] These studies show the promise of tissue engineered skeletal muscle constructs as therapeutic neovascularization devices; however, the ability of these constructs to generate favorable results with respect to functional outcomes and arteriogenesis remains to be determined.

## 2.3 Injectable Therapeutics: Systemic Injection

In comparison to the relatively high number of implantable constructs, scaffolds, and other biodegradable polymer delivery devices that are being developed for stimulating therapeutic neovascularization, there are few bioengineering approaches under development that utilize the delivery of growth factor bearing vehicles via injection into the systemic circulation. From a clinical perspective, the key advantage to such a strategy is that the intervention itself (ie. intravenous or intra-arterial injection) may be considerably less invasive than a surgical procedure to implant a construct or device for triggering neovascularization. However, there are also a number of potential hurdles that a strategy utilizing an injectable delivery device must overcome, and these hurdles are likely the reason why fewer strategies involving injectable delivery vehicles are under development. For instance, in most cases, one would want the intravascular therapy to be delivered to the ischemic tissue which, by definition, has limited or no blood flow. This flow deficit can, in turn, limit the effective concentration of the agent in the ischemic region, thereby reducing the desired therapeutic effect. Furthermore, once these vehicles have been convected to the intended site of action, they must stop and remain long enough to generate beneficial remodeling. Given these limitations, such intravascular delivery vehicles must, in all likelihood, undergo some type of targeting to the region of interest that will enhance both their local concentration and duration of action. This targeting could be done actively through adhesion to endothelial antigens that are expressed in a site-specific manner. However, while considerable effort and progress has been made in identifying tumor specific antigens for the targeting of anti-angiogenic therapies, the identification of endothelial-specific markers expressed in ischemic tissue or tissues surrounding zones of ischemia has yet to lead to targeted therapies for stimulating neovascularization.

### 2.3.1 Gelatin Hydrogel Microspheres

Alternatively, as shown by Hosaka et al.[26], targeting may be done passively via local intravascular injection of vehicles that become physically obstructed in the microcirculation. In their work on the use of gelatin hydrogel microspheres for bFGF delivery to ischemic hindlimb, Hosaka et al.[26] demonstrated that, as expected, 29 μm microspheres accumulated in ischemic hindlimb at a far higher level than 10 μm microspheres. Of course, when using such a strategy, one must be concerned with the potential for these obstructions to further deteriorate tissue perfusion. In their study, Hosaka et al.[26] took this consideration and determined that arterially injected 29 μm micropsheres had no immediate deleterious effects on regional blood flow in the hindlimb. However, larger diameter microspheres (ie. 59 and 75 μm diameter) did create a significant drop in regional blood flow. Chronically, when compared to hindlimbs treated with PBS bearing gelatin hydrogel microspheres, Hosaka et al.[26] measured positive changes in a number of important parameters, including capillary density, angiographic score, collateral conductance, and blood flow in 29 μm diameter bFGF micropshere treated hindlimbs. Importantly, these values were also observed to be above those measured in hindlimbs treated with intramuscular bFGF injection, thereby highlighting the added benefit of applying a sustained dose of growth factor over an extended period of time.

### 2.3.2 Ultrasound Microbubble Based Systems

Another emerging bioengineering technology that utilizes injectable delivery vehicles and may be well-suited to overcoming the challenges of intravascular delivery employs the interaction of ultrasound with intravascular contrast agent microbubbles. Microbubble (2-10 μm diameter) contrast agents typically consist of an engineered shell surrounding air or a high-molecular weight gas. Several shell compositions have been developed, including albumin, lipid, surfactant, complex sugars, and polymers.[27] As implied by their name, the original intended use of microbubble contrast agents was to enhance contrast during ultrasound imaging, and microbubble contrast agents have been shown to yield significant opacification of myocardial borders, the liver, the kidney, and tumors.[28-30] However, several studies also indicate that ultrasonic contrast agents may be ideally suited for targeted drug/gene delivery[31-35], as well as the direct creation of arteriogenesis.[36,37]

There are two basic premises that underlie the use of ultrasound-microbubble interactions for targeted drug/gene delivery and the creation of neovascularization. The first is that, under certain user-defined conditions, contrast agent microbubbles can be induced to go into a mode of nonlinear cavitation that results in their destruction.

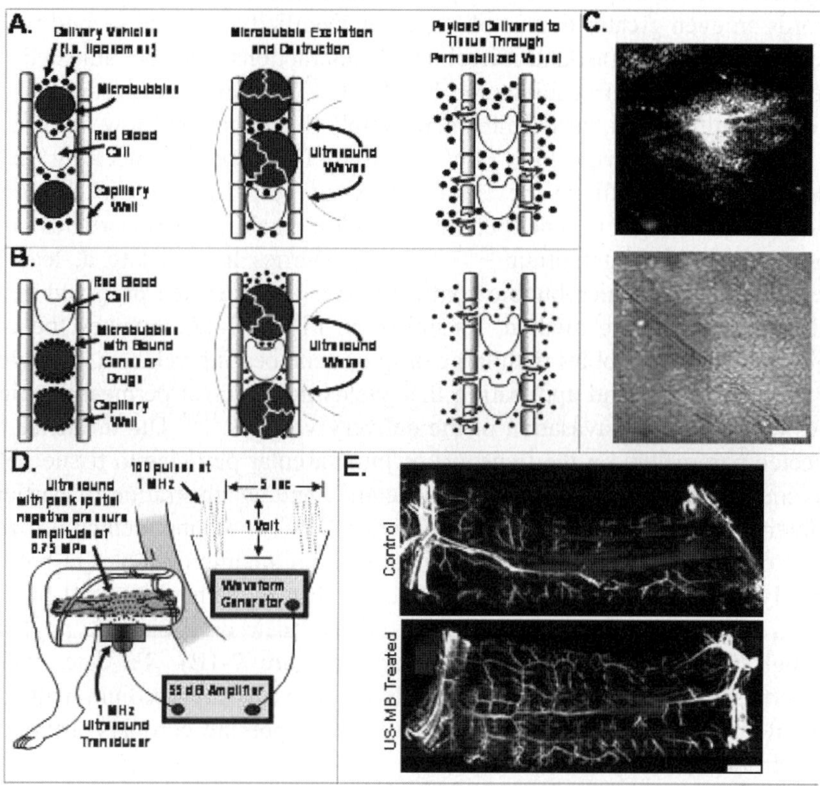


*Figure 7-1.* Ultrasound-microbubble interactions for targeted therapeutic applications. A: Schematic illustration of how targeted gene or drug delivery may be achieved with ultrasonic microbubble contrast agents. Microbubbles and delivery vehicles are co-injected, ultrasound is applied to the region of interest, microbubbles are destroyed and vessels are permeabilized, and delivery vehicles are delivered to tissue. B: Alternatively, delivery vehicles or biological agents are bound to the bubbles and then injected, ultrasound is applied to the region of interest, the payload is released form the bubble, vessels are permeabilized, and the payload is delivered. C, top: 100 nm fluorescent particles that have been delivered from the intravascular space to muscle tissue via a capillary pore generated by microbubble destruction with ultrasound. C, bottom: Same region of muscle shown with transillumination. Bar indicates 100 μm. D: Schematic illustration of the experimental apparatus used to generate arteriogenesis in rat or mouse gracilis muscle via ultrasound-microbubble interactions. E: Whole mount gracilis muscles that have been labeled for SM α-actin 14 days after receiving ultrasound-microbubble or sham control treatment. Note the enhanced density of SM α-actin positive vessels with ultrasound-microbubble treatment. Bar indicates 1 mm.

The ultrasound beam can be targeted or tailored according to user specifications, thereby limiting the release of the drugs/genes of interest to the ultrasound targeted region. Some microbubbles have also been engineered to bind specifically to pre-selected endothelial antigens, and this affords an even greater degree of targeting specificity.[38-42] The second basic premise is that ultrasound-microbubble interactions can be adjusted to produce capillary permeabilization bioeffects.[43-47] Typically, in comparison to clinical ultrasound, these adjustments imply that either very low frequency (1 MHz or less) or very high power is used. The pores produced by these interactions can facilitate the extravasation of very large (up to 500 nm) intravascular drug or gene bearing vehicles, such as nanoparticles and liposomes, to the interstitium.[33,34] These properties have led to at least 2 methods for using microbubble contrast agents as a targeted drug and gene delivery system. One method, illustrated in Figure 7-1A, invokes the co-injection of microbubbles with large drug or gene bearing vehicles, followed by targeted ultrasound application that yields microvessel permeabilization sites that permit extravasation of the delivery vehicles.[33,34] The influence of selected parameters on the transport of intravascular particles to tissue with this method has now been studied in detail[34], and an illustration of particle delivery with this method is shown in Figure 7-1C. A second method utilizes drugs or genes that are either encapsulated within the microbubbles or bound directly to the microbubble shells. Once destroyed with ultrasound in the region of interest, these microbubbles release their drug or gene payload through newly generated capillary pores (Figure 7-1B). To date, these properties have been harnessed to deliver genes to myocardium with an adenovirus[31] and deliver plasmid DNA to both porcine coronary arteries[48] and carotid arteries.[49]

Progress has also been made in the use of contrast agent microbubbles for generating beneficial neovascularization.[32] Microbubbles conjugated to VEGF have been injected intravenously and subsequently destroyed in rat myocardium with ultrasound. Following treatment, VEGF was primarily deposited in the endothelium of intramyocardial arterioles, and increases in both endothelial and smooth muscle cell number were observed, suggesting that the deposited VEGF had induced vascular cell proliferation. However, even though cell proliferation increased, changes in vascular density were not measured. More recently, Kondo et al.[35] examined the use of ultrasound contrast agents for the non-viral delivery of a plasmid encoding hepatocyte growth factor (HGF), an angiogenic factor, to infarcted myocardium. In this study, intermittent ultrasound was applied for 2 minutes as plasmid was infused into the left ventricle in the presence of contrast agent microbubbles. It is important to note, however, that HGF plasmid and the contrast agent was not attached to the microbubbles. The results from this work indicated that HGF protein was expressed only with ultrasound-microbubble treatment, compared to controls in which only ultrasound was applied or empty plasmid was injected. Furthermore, the authors noted significant reductions in left ventricular volume and scar size that were accompanied by increases in both capillary and arteriole density. This was the first study to demonstrate that ultrasound-microbubble interactions can be used to deliver a pro-angiogenic gene that stimulates a therapeutic response that, in turn, ameliorates complication associated with myocardial infarction.

Transmyocardial laser revascularization (TMLR) procedures may be used to initiate an inflammatory response, which in turn, elicits neovascularization.[50-55] Given these observations, Song et al.[36] hypothesized that bioeffects created by high power/low frequency ultrasound could generate neovascularization the same manner as seen with TMLR. This hypothesis has now been tested in two separate studies. In the first[36], ultrasound-microbubble induced bioeffects were created in normal rat gracilis muscles by applying pulsed 1 MHz ultrasound following intravenous microbubble injection using the experimental apparatus outlined in Figure 7-1D. After verifying that ultrasound application had elicited capillary poration, animals were allowed to recover. Muscles were examined 3 to 28 days later for changes in arteriolar density and hyperemia blood flow. Consistent with the hypothesis that bioeffects were creating beneficial arteriolar remodeling, we observed a 65% increase in SM α-actin positive vessels per fiber over sham control after 1 or 2 weeks. Importantly, this increase was observed across multiple diameter ranges and was accompanied by a 60% increase in hyperemia nutrient blood flow to the gracilis muscle at 2 weeks. Ultrasound application alone yielded no increase in arteriogenesis or hyperemia blood flow indicating a requirement for microbubbles to be present.

While these first studies demonstrated the feasibility of using ultrasound-microbubble interactions for stimulating arteriogenesis and increased hyperemia blood flow in skeletal muscle, a key question was whether such a treatment could restore hyperemia blood flow in muscle affected by arterial occlusion. To this end, we examined arteriogenesis, angiogenesis, and hyperemia blood flow following the creation of bioeffects via ultrasound-microbubble interactions when one of the two main feeding arterioles to rat gracilis muscle was occluded.[37] Using essentially the same assays as Song et al.[35], microvascular remodeling and hyperemia blood flow were assessed at 1, 2, and 4 weeks after arterial occlusion and treatment. In this study, compared to sham controls, statistically significant 30-40% increases in hyperemia flow were measured at all timepoints in the region of the muscle closest to the arterial occlusion. Moreover, hyperemia flow was within 5% of normal levels in this region of muscle at 4 weeks. Interestingly, arteriogenesis was increased throughout the muscle, but was especially evident in the muscle region around the patent feed artery at 2 and 4 weeks after treatment, consistent with the hypothesis that flow restoration was due to an increase in the number and caliber of smooth muscle coated collateral vessels connecting the patent feed artery to the region of low perfusion. Examples of whole-mounted ultrasound-microbubble treated and contra-lateral control muscles gracilis muscles that have been labeled for SM α-actin are presented in Figure 7-1E. Finally, while an increase in capillary density was observed shortly after treatment, this response was reduced at day 28 when hyperemia flow was restored, thereby suggesting that capillary growth had a minimal effect on flow.

Looking forward, there are several challenges that remain in determining whether the use of ultrasound-microbubble mediated neovascularization can become a viable clinical alternative to conventional surgical. For example, the mechanisms through which ultrasound-microbubble generated bioeffects create arteriogenesis must be explored in more detail. As with any new technology, safety issues must also be addressed. Moreover, translation of this work to humans could be particularly difficult in light of the fact that bioeffect creation appears to be at least somewhat species dependent.[56] Finally, it will be interesting to determine whether the baseline arteriogenesis response can be amplified and sustained through the additional delivery of growth factor protein and gene bearing particles and liposomes to ultrasound targeted tissue. Indeed, unless ultrasound-microbubble mediated procedures can be performed multiple times, this stimulus itself is brief and the effects possibly transient. For this reason, combining the original stimulus with the slow and sustained of a growth factor may eventually represent a more advanced strategy.

## 2.4 Injectable Therapeutics: Intramuscular and Intramyocardial Injection

For certain applications, therapeutic strategies that utilize intramuscular or intramyocardial injection of growth factor delivery vehicles may be an appropriate solution. In the scenario of limb claudication, for example, regions surrounding the occluded femoral artery may be easily amenable to intramuscular injection. Many of these injectable systems are based on the same polymers that were described previously when discussing implantable therapeutics. For example, Cleland et al.[57] used poly(lactide-co-glycolide) (PLG) to generate a bead delivery system for recombinant (rh)EGF, with the explicit goal of using such a system for the delivery of therapeutic growth factors via intramuscular or intramyocardial injection. In their study, rhVEGF containing PLG microspheres produced a dose-dependent angiogenic response in a rat cornea angiogenesis model, indicating the promise of using PLG bead delivery systems for stimulating therapeutic angiogenesis.

Alginate has also been "molded" into an injectable (ie. bead) form to support the delivery of angiogenic growth factors. To this end, Downs et al.[58] adapted an alginate based system for the entrapment of tumor cells. TGF-α, bFGF, and EGF were entrapped in alginate during the gelation process, during which small injectable beads were formed. After demonstrating the release kinetics and bioactivity *in vitro*, the authors injected these alginate beads subcutaneously into balb/C mice and, using red blood cell pooling as an angiogenic metric, demonstrated a dose-dependent neovascularization response with each of the three factors. More recently, in response to the relatively poor release profiles afforded by alginate in other earlier studies, Gu et al.[59] examined how modulation of a number of factors, including electrostatic interaction, growth factor (VEGF) concentration, and $CaCl_2$ concentration during the gelation process alters the rate and duration of VEGF delivery from alginate beads, concluding that these parameters can be successfully manipulated to generate constant and sustained (14 day) release of VEGF. These modifications were not, however, tested *in vivo* with respect to neovascularization.

Finally, promising advances in the use of gelatin hydrogels as growth factor and gene delivery devices following intramyocardial and intramuscular injection have also been recently made. Sakakibara et al.[60] examined the potential benefit of injecting bFGF in gelatin hydrogel beads to ischemic myocardium before performing cardiomyocyte transplantation. Here, echocardiography revealed that injected bFGF gelatin hydrogel beads, both alone and in conjunction with cardiomyocyte transplantation, improved ventricular shortening. Moreover, bFGF bead pretreatment also improved left ventricular maximum time-varying elastance when compared to all other interventions. Histologic analysis demonstrated 2-3 fold increases in capillary density in the scar and peri infarct regions with bFGF gelatin hydrogel beads, as well as an increase in cell survival amongst transplanted myocytes, supporting the hypothesis that prevascularization via the controlled release of bFGF significantly enhanced the efficacy of the transplant. Gelatin hydrogels have also been used as injectable delivery devices for gene therapy with plasmid FGF4.[61] In this study, performed on ischemic mouse and rabbit hindlimb, gelatin hydrogels were shown to improve transfection efficiency. In comparison to controls (ie. naked DNA injection or gelatin hydrogels bearing lacZ reporter gene), FGF4 gene hydrogels elicited significant improvements in necrosis and hyperemia blood flow. These improvements were accompanied by an increase in angiographic score during vasodialtion, indicative of beneficial lumenal remodeling of the arterioles and arteries.

## 3. NEOVASCULARIZATION IN TISSUE ENGINEERING

Broadly defined, the term "tissue engineering" refers to the use of engineering principles for the replacement, repair, and regeneration of living tissues. These goals of tissue engineering are typically addressed through the rational combination of selected cells, biological and chemical factors, and biomaterials, and several new advances in the field have revolved around the generation of so-called "tissue engineered constructs". These constructs typically consist of two primary ingredients - a scaffold and cells. The scaffolds are usually porous 3-dimensional biomaterials that may be derived from either synthetic or natural polymers. In most instances, the scaffold is designed to degrade after the construct has been implanted and subsequently invested with tissue from the host organism. The cellular component(s) of the construct may consist of one cell type, multiple cell types, native cells, and/or genetically engineered cells. Cells may be, or may be derived from, stem cells, or they may be tissue specific cells, or even vascular cells. Tissue constructs are often generated *in vitro* through the addition of the adherent cells to the 3-dimensional porous polymer scaffold *in vitro*. Following an appropriate amount of time *in vitro*, constructs are implanted to the defect site as either a substitute for the missing or defective tissue or as a foundation on which new host tissue can incorporate.

While it is true that approaches utilizing tissue engineered constructs continue to hold great promise for the future of health care, recent progress, for the most part, has been slow. Amongst many of the remaining challenges for tissue engineering, one critical problem has been the inability of some tissue engineered constructs to become functionally vascularized. The requirement for construct vascularization is not necessarily absolute, and many tissue constructs survive without incorporating host blood vessels; however, constructs of considerable thickness and/or containing cells with relatively high metabolic activity will often fail without a sufficient vasculature.

*Table 7-2*. Bioengineered Scaffolds that Facilitate Vessel In-Growth.

| **Scoffold Comosition** | **Impolntation Site** | **Reference** |
|---|---|---|
| small intestinal submucosa | bladder | Kropp et al.(1995)[67] |
| acellular dermis | skin | Wainwright et al.(1995)[68] |
| candaveric fascio | ligament | Curtis et al.(1985)[69] |
| bladder acelluar matrix | bladder | Reddy et al.(2000)[70] |
| PLG bearing VEGF and PDGF | subcutis | Richardson et al.(2001)[71] |
| osteoinductive PLG bearing VEGF | tested in vitro | Murphy et al.(2000)[72] |
| fibrin bearing aFGF | ear | Pandit et al.(1998)[73] |
| alginate hydrogel bearing FGF-2 | subcutis | Lee et al.(2000)[51] |
| fibrin impregnated with FGF-2 | subcutis | Fourier and Doillon (1996)[74] |
| polyurethane and gelatin bearing VEGF | vascular graft | Masuda et al.(1991)[75] |
| matrix modified ePTFE | subcutis | Kidd et al.(2002)[76] |
| matrix modified ePTFE | subcutis | Kidd and Williams (2004)[77] |

PLG=poly(lactied-co-glycolide), VEGF=vascular endothelial growth factor, PDGF=platelet deriyed growth factor, aFGF=acidic fibroblast growth factor, ePTFE-expanded polyterafluoroethlene.

Upon implantation, biomaterials elicit a foreign-body response that includes the formation of a multi-layer capsule around the implant itself. As reviewed by Sieminski and Gooch[62], it is well-known that the vascularization of this capsule surrounding many biomaterials, as well as the biomaterial itself, is largely dependent on biomaterial microarchitecture. Both Padera and Colton[63] and Brauker et al.[64] studied the effect of biomaterial porosity on vascularization and found biomaterials with pore sizes of approximately 10 μm are amenable to generating vascularized capsules. It was also shown that smaller pore sizes generate avascular capsules. Further comprehensive evidence that porosity has a substantial effect on capsule vascularization and permeability was presented by Sharkawy et al.[65] In contrast, the chemical composition of the biomaterial has comparatively little effect on vascularization of the surrounding capsular layers.

While these studies indicated that biomaterial microarchitecture can be tuned to generate capsule vascularization that could perhaps support the metabolic needs of cells within the biomaterial implant, it is important to note that many new generation tissue engineered constructs are designed to mimic and replace specific tissues. Because these tissues may not necessarily contain a matrix component with a porosity that is advantageous for vascularization, it is clearly desirable to develop strategies for eliciting vascularization that eliminate these constraints on scaffold porosity altogether. Moreover, it is clear that coaxing vessels beyond the surrounding capsule and into the construct itself would be advantageous. To date, such strategies for eliciting deep vessel growth often involve the addition of specific biological signals, such as diffusible chemotactic growth factors, or matrix modifications that facilitate cell migration and adhesion. Alternatively, the cells within the tissue constructs may be chosen or manipulated to recruit host vessels into the construct. In the following sections, we will review many of these novel strategies and recent advances in eliciting vascularization of bioengineered tissue constructs. The first three sections (ie. *3.1, 3.2,* and *3.3*) focus purely on scaffolds that either naturally, or through added biological signals, stimulate vessel in-growth. Table 7-2 summarizes these studies. In the fourth section (*3.4*), vascularization of complete tissue constructs, including those that include endothelial cells, is examined. Finally, in the fifth section (*3.5*), tissue constructs in which microvessels are formed *in vitro*, before construct implantation, are discussed. Table 7-3 summarizes those studies presented in sections *3.4* and *3.5*.

*Table 7-3.* Bioengineered Tissue Constructs that Facilitate Vessel In-Growth.

| Construct Composition | Implantation Site | Reference |
|---|---|---|
| PLA housing Bcl-2 modified HDMEC | subcutis | Nor et al. (1999)[78] |
| collagen-fibronectin gels with Bcl-2 modified HUVEC | abdominal wall | Enis et al. (2005)[80] |
| telomerized HDMEC in Matrigel | subcutis | Yang et al. (2001)[79] |
| PLA/PLG with ESC derived endothelial cells | subcutis | Levenbury et al. (2002)[82] |
| PLG/VEGF scaffold with HMVEC | subcutis | Peters et al. (2002)[12] |
| PEGT/PBT scaffold with fibroblasts | full-thickness wounds | Wang et al. (2004)[83] |
| artificial dermis with bFGF gelatin microspheres | full-thickness skin defects | Kawai et al. (2000)[84] |
| PLA/bFGF discs with hepatocytes | small bowel mesentery | Lee et al. (2002)[85] |
| PLG/VEGF with hepatocytes | subcutis | Smith et al. (2004)[13] |
| PLG sheets with intestinal organoids | subcutis | Gardner-Thorpe et al. (2003)[86] |
| keratinocytes and fibroblasts on collagen | in vitro | Black et al. (1998)[87] |
| keratinocytes and fibroblasts on collagen | in vitro | Hudon et al. (2003)[88] |
| adipose and endothelial cells on fibrin | in vitro | Frerich et al. (2001)[69] |
| endothelial cells on modified fibrin | in vitro | Hall et al. (2001)[50] |
| endothelial cells on fibronectin | in vitro | Pome et al. (2004)[51] |
| microvessel fragments in collagen gels | | Shepherd et al. (2004)[52] |

PAL=polylatic acid,HDMEC=human dermal microvascular endothelial cells, HUVEC=human umbilical vein endothelial cells, PLG-polycolic acid, ESC=embryonic steam cell, HMVEC=human mjciovascular endothelial cells, PEGT=polyethyleneglycol terephthalate, PBT=polybutylene terephthalate, b FGF=basic fibroblast growth factor, VEGF=vascular endothlial endothelial growth factor

## 3.1 Scaffolds from Naturally Occurring Matrix Materials

Naturally occurring matrix materials offer an attractive choice for tissue engineering scaffolds. These scaffolds tend to interface well with host tissues and both their microarchitecture and biochemical composition are naturally amenable to supporting neovascularization. Indeed, as discussed in the review by Hodde et al.[66], naturally occurring extracellular matrices often contain bound growth factors, such as bFGF and TGFβ, that may serve as chemotactic agents for new vessels and assist with the maturation of recently formed vessels within the scaffold. Hodde et al.[66] subdivided naturally occurring matrices into several categories, including small intestinal submucosa, acellular dermis, cadaveric fascia and bladder acellular matrix. Each of these naturally occurring scaffold materials has been shown to support neovascularization *in vivo*. Kropp et al.[67] reported that, when used as a bladder patch, small intestinal submucosa implants exhibited significant new capillary growth in conjunction with restoration of bladder tissue. Wainwright et al.[68] demonstrated that acellular dermis becomes well-vascularized and this may be a significant reason why acellular dermis holds promise as a treatment for full thickness burns. Cadaveric fascia, which has been tested as a ligament graft, becomes vascularized after implantation[69], and bladder acellular matrix grafts, intended to restore bladder function after implantation, facilitate the incorporation new microvessels.[70]

More recently, Anderson et al.[71] reported the presence of a naturally occurring matrix scaffold that directly supports the adhesion and migration of capillary sprout endothelial cells and pericytes. Specifically, the endogenous elastic fiber matrix of rat mesenteric connective tissue microguides the leading endothelial cells of capillary sprouts, such that these migrating cells essentially follow the longitudinal axis of individual elastic fibers. An example of a capillary sprout leading endothelial cell migrating along an elastic fiber in rat mesentery is shown in Figure 7-2. Because elastic fibers may be comprised of many different matrix proteins, including fibrillins I and II, elastin, microfibril associated glycoproteins I and II, and latent TGFβ-binding protein, and each of these proteins may contain several motifs for integrin and non-integrin mediated cell adhesion, it is difficult to ascertain which sites are being utilized. However, even with these caveats, these results do lend support to the hypothesis that tissue constructs, based on fiber scaffolds with appropriate dimensions and biological cues for endothelial cell migration, may be useful for stimulating microvessel in-growth. Moreover, while biomimetic tissue constructs based on mesenteric connective tissue have yet to be tested *in vivo*, these results may hold important clues for how biomimetic tissue constructs which are capable of microguiding capillary sprouting may be designed.

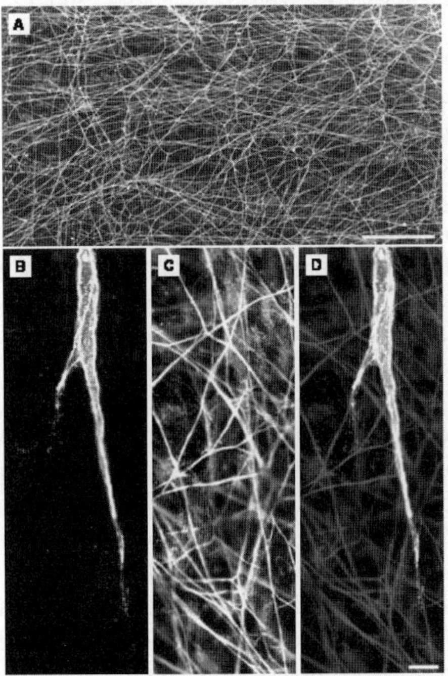


*Figure 7-2.* Confocal microscopy images depicting capillary sprout endothelial cells and the elastic fiber network in rat mesenteric connective tissue. A: Low magnification image of elastic fiber network revealed by immunochemistry with an anti-elastin antibody. B: Capillary sprout leading endothelial cell immunochemically labeled with an antibody to the endothelial specific marker CD31. C: Elastic fibers in the same region of tissue shown in B. D: Merge of panels B and C illustrating that the capillary sprout leading endothelial cell is in contact, and aligned, with an underlying elastic fiber. Bars indicate 200 μm (A) and 10 μm (D).

## 3.2 Scaffolds with Incorporated Soluble Growth Factors

To date, strategies for eliciting the directed growth of new vessels into a porous biomaterial that are based on the incorporation of soluble growth factors into synthetic or natural polymer scaffolds have shown promise. As these bioengineered scaffolds degrade or dissolve, the soluble factors are released into the interstitium, thereby creating a growth factor concentration gradient that stimulates the migration of endothelial and smooth muscle cells into the scaffold. An illustration of this process is presented in Figure 7-3A. Many of the synthetic polymers that have been engineered to release growth factors during degradation were previously introduced when discussing controlled release systems for eliciting therapeutic neovascularization. For instance, while Richardson et al.[14] demonstrated that their PLG based scaffold system for the sequential release of VEGF and PDGF generated neovascularization in surrounding tissue, they also demonstrated a substantial increase in vessel incorporation into the scaffold itself. This study was preceded by a study by from the same group[11] wherein the ability of mineralized osteoconducive PLG matrices to become vascularized was explored using VEGF as the angiogenic stimulus. Although these matrices were not tested *in vivo*, the controlled release of biologically active VEGF was confirmed via the ability of released VEGF to stimulate the proliferation of human dermal microvascular endothelial cells. Finally, when used in conjunction with a collagen coating, the delivery of acidic FGF (HBGF-1) from PTFE fibers creates a neovascularization response that supports the survival of organoids *in vivo* that are derived from transplanted rat hepatocytes.[72]

Scaffolds comprised of naturally derived materials have also been modified to deliver soluble growth factors and enhance vascularization. As early as 1988, Thompson et al.[6] reported that gelatin sponges that had been modified to contain aFGF generated a much higher vascularization response than control sponges upon implantation *in vivo*. More recently, Pandit et al.[73] delivered aFGF from modified fibrin scaffolds to assist with healing in a rabbit ear ulcer model. They described a correlation between the aFGF dose and the healing response and noted that aFGF delivery significantly enhanced neovascularization. Also, in a study that was previously described, Lee et al.[9] have shown that mechanical compression of alginate hydrogels loaded with VEGF generated a substantial increase in blood vessel density in the granulation tissue surrounding the implant. While vessels were not reported within the alginate hydrogel itself, in many applications, the growth of vessels in the regions around certain biomaterials and constructs may be sufficient to support the metabolic requirements of the cells within.

Finally, some soluble growth factor release systems have also been generated via the combination of natural and synthetic polymer materials. For example, by adsorbing a fibrin matrix that was impregnated with FGF-2 (basic FGF) and an endothelial cell growth supplement (ECGS) to a polyester prosthesis mesh, Fournier and Doillon[74] generated a composite scaffold that was designed to stimulate vascularization. Although most of the FGF-2 was released from this composite scaffold within the first day, both FGF-2 and ECGS impregnated composites generated increased microvessel density. Similarly, Masuda et al.[75] combined polyurethane and gelatin to form a vascular graft with a base composite material that permitted the controlled release of VEGF. In grafts with suitable pore sizes (100 μm), those releasing VEGF exhibited significant new capillary infiltration.

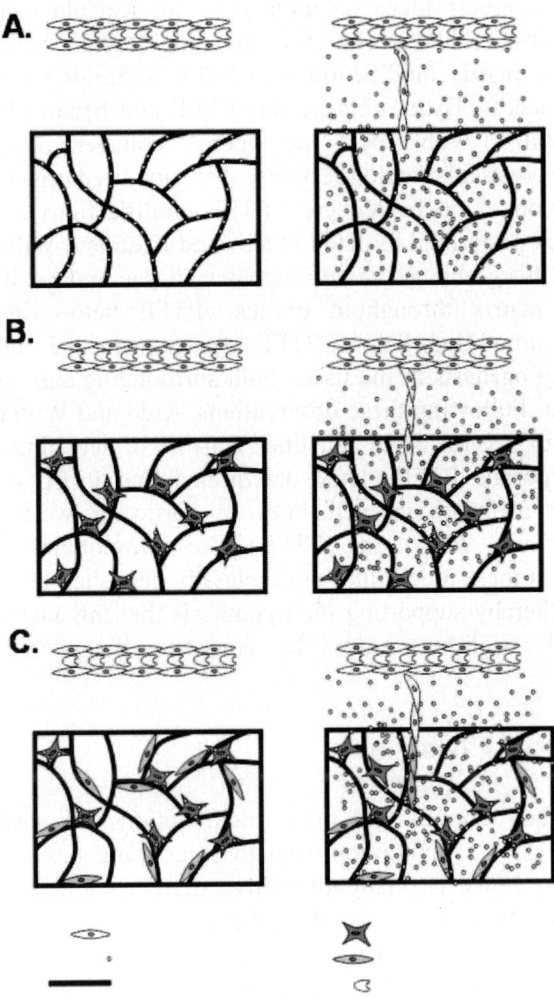


*Figure 7-3.* Schematic illustrations depicting multiple strategies for eliciting vessel growth into bioengineered polymer scaffolds and tissue constructs. A, left: The incorporation of soluble growth factor(s) into a biodegradable scaffold. A, right: As the scaffold degrades, growth factors are released, thereby creating a chemotactic gradient that recruits host vasculature into the scaffold. B, left: Tissue-specific cells adhered to scaffolds before implantation. B, right: The tissue specific cells adhered to scaffolds secrete growth factors, either naturally or through genetic manipulations, that recruit host vessels into the construct. C, left: Endothelial cells may be incorporated into a tissue construct. C, right: The implanted endothelial cells provide an immediate source of cells for chimeric vessel formation.

## 3.3 Scaffolds with Matrix Modifications for Cell Adhesion

Other cutting edge strategies for supporting the vascularization of biomaterial constructs focus on modifying the scaffold material to better support the adhesion and migration of invading vascular cells. In particular, pro-angiogenic matrix modifications to PTFE scaffolds have been popular and shown success. By coating porous PTFE and heparin binding growth factor-1 (HBGF-1) with type I and type IV collagen, Thompson et al.[72] stimulated a neovascularization response that supported organoids created by the implantation of rat hepatocytes. PTFE modifications for angiogenesis have also been performed by Kidd et al.[76] and Kidd and Williams[77]. In their earlier study, this group used a tumorigenic cell line to deposit an angiogenic extracellular matrix throughout porous ePTFE before implantation. In contrast to unmodified PTFE, PTFE coated with the modified matrix stimulated angiogenesis in the tissue both surrounding and within the pores of the implant. Following these observations, Kidd and Williams[77] extended their studies to several other cell lines capable of secreting an angiogenic matrix onto porous PTFE. They determined that PTFE matrix coatings generated by 3 individual cell lines in particular were able to elicit neovascularization within the PTFE pores. Importantly, it was also determined that these 3 cell lines preferentially deposited laminin-5 onto the PTFE discs, thereby supporting the hypothesis that this matrix protein may be particularly useful as a tool for amending biomaterial implants for angiogenesis.

## 3.4 Tissue Constructs

As described above, biological modifications to scaffold polymer materials represent a viable approach to stimulating tissue construct neo-vascularization; however, many alternative strategies utilize the cells within the constructs themselves as the primary agents for triggering and maintaining new vessel growth. In general, there are two ways this may be done. The first method involves the incorporation of vascular cells (typically endothelial cells) into the construct, with the intent that these incorporated vascular cells will become a source for neovessel formation. Endothelial cells may either be randomly distributed throughout a matrix, or as described in section *3.5* which focuses on microvessel implantation, they may be conditioned to form endothelial tubes in the matrix before implantation. These general strategies are illustrated in Figure 7-3B. The second method, which is presented schematically in Figure 7-3C, relies on the ability of tissue specific cells within the construct to generate pro-angiogenic signals that recruit vessels from the host organism into the construct.

To date, most angiogenesis studies involving the implantation of endothelial cell bearing matrices have been done without any other cell type present in the matrix before implantation. In other words, these are not necessarily complete tissue engineered constructs and therefore, in the context of tissue engineering, much of this work is currently at a "proof-of-principle" stage wherein the ability of these simpler implants to elicit neovascularization is being tested. Even with these caveats, these approaches have yielded many promising results, and as model systems for studying angiogenesis, they have often contributed considerably to our basic understanding of angiogenesis and vessel maturation. A number of these key studies have involved genetic manipulations to enhance endothelial cell survival and angiogenesis.[78-80] For example, Nör et al.[78] implanted poly-lactic acid sponges bearing human dermal microvascular endothelial cells that overexpressed the anti-apoptotic protein Bcl-2 into SCID mice subcutaneously. In comparison to sponges bearing normal human dermal endothelial cells, sponges with Bcl-2 transfected endothelial cells exhibited an approximately 5-fold greater microvessel density and a significant decrease in apoptosis. This same group later demonstrated that untransduced human endothelial cells, implanted in the same manner, organize into empty tube structures before anastomosing with the host vasculature and becoming perfused.[81] More recently, Enis et al.[80] used a similar model system wherein Bcl-2 overexpressing human umbilical vein endothelial cells were implanted into SCID mice using collagen-fibronectin gels. In addition to observing a stable vasculature containing mature arterioles and venules within the gel implants, the authors noted that the vessels within the gel implants were chimeras of human endothelial cells and mouse perivascular cells. Yang et al.[79] have also used genetic manipulation to enhance endothelial survival, showing that human dermal endothelial cells transduced with human telomerase reverse transcriptase anastomose with SCID mouse vessels upon implantation with matrigel. While vasculature derived from primary dermal endothelial cells was shown to regress with time, the vasculature from telomerized endothelial cells remained patent for several weeks. Levenburg et al.[82] induced the differentiation of human embryonic stem cells into endothelial cells and subsequently implanted these cells into SCID mice using PLLA/PLGA scaffolds. These implants exhibited vessels comprised of human endothlelial cells that contained mouse red blood cells, indicating that the human vessels had inosculated with the host vasculature. Finally, using an approach that combined human microvascular endothelial cells with a controlled release PLG scaffold housing VEGF, Peters et al.[12] reported that, upon implantation into SCID mice, the VEGF scaffold supported a significant increase in human microvessel density when compared to control scaffolds without growth factor.

Other studies that have utilized tissue specific cells, but not endothelial cells, for stimulating neovascularization, are often based on tissue constructs for dermal replacement and regeneration. As previously noted, three dimensional fibroblast cultures (3DFCs) have been tested as a means for generating therapeutic angiogenesis. However, these constructs may also be useful as a treatment for diabetic foot ulcers, and it has been shown that they secrete VEGF and stimulate endothelial cell proliferation and vessel growth.[22] The vascularization of dermal equivalents generated on poly-ethyleneglycol terephthalate (PEGT)-polybutylene terephthalate (PBT) multiblock copolymer scaffolds using fibroblasts from multiple tissues was studied by Wang et al.[83] Here, in comparison to acellular scaffolds, constructs cultured with fibroblasts from multiple sources generated significant angiogenesis. To enhance vascularization of an artificial dermis construct comprised of a collagen sponge layer and a silicon layer, Kawai et al.[84] added controlled release gelatin microspheres to the dermis and observed that the number of capillaries within the construct increased in an approximately dose-dependent manner.

Tissue constructs designed for hepatocyte transplant and intestinal replacement have also been amended to increase directed neovascularization. Hepatocytes, in particular, have high metabolic activity that must be served with an adequate vascular supply following implantation. To this end, Lee et al.[85] transplanted hepatocytes on poly-lactic acid discs coated with bFGF or a control solution and determined that bFGF supported a substantial enhancement of hepatocyte engraftment. Using a similar approach, Smith et al.[13] recently employed the now well-developed PLG/VEGF delivery scaffold as a support for transplanted hepatocytes. From these studies, it was found that the VEGF scaffold enhanced vascularization and provided a significant increase in hepatocyte survival when compared to cells transplanted on scaffolds without growth factor. With regard to tissue-engineered small intestine generated by seeding polyglycolic acid sheets with intestinal organoids, Gardner-Thorpe et al.[86] noted that the muscle and mucosal layers of the tissue generated from these grafts exhibit capillary densities that are essentially equivalent to normal adult tissue.

## 3.5 Microvessels Grown in Constructs *In Vitro*

The incorporation of endothelial cells into constructs before implantation has been discussed previously. In this final section, a specific subset of endothelial transplantation studies will be addressed. Specifically, those strategies in which microvessel organization is stimulated *in vitro* will be presented. In essence, the goal here is to prevascularize a construct before implantation with the hope of eventually generating chimeric vessels in which tubular vessel-like structures, that were generated *in vitro*, inosculate with the host vasculature. The ability of such chimeric vessels to be generated was documented above; however, by organizing vessels within the construct *in vitro* before implantation, an additional level of control over construct vascularization may eventually be possible.

One such example of preformed microvessel like structures in a tissue construct *in vitro* was presented by Black et al.[87] Multi-layered skin equivalent constructs comprised of keratinocytes and fibroblasts on a collagen biopolymer were endothelialized by the addition of human umbilical vein cells. Using immunochemistry for selected endothelial markers and laminin, as well as general electron microscopy, Black et al.[87] described the formation of tubular structures after 15 and 31 days in culture that presented intercellular junctions between endothelial cells. Hudon et al.[88] then followed this study with additional work aimed at understanding how selected pro- and anti-angiogenic factors influence tube formation in these skin equivalents. The addition of bFGF and VGEF to the constructs enhanced *in vitro* tube density, while progesterone reduced tube density. Moreover, confocal microscopy confirmed the formation of endothelialized lumens in these skin equivalents. Although the ability of these pre-formed vessels to inosculate with host circulation has not been presented, the ability to adjust the density of pre-formed microvessels *in vitro* through pharmacological interventions illustrates how vessel networks may be engineered *in vitro* before implantation.

Other strategies to manipulate microvessel development *in vitro* employ selected biomolecular interventions of the matrix or supporting scaffold. In a construct composed of adipose cells and endothelium, Frerich et al.[89] incorporated a fibrin-microcarrier scaffold and observed the formation of microvessel network structures that were responsive to VEGF application. A fibrin based matrix was also used as a base material in a study by Hall et al.[90] Here, it was shown that alterations in pH triggered the formation of longitudinally oriented fibrils that supported *in vitro* angiogenesis. Furthermore, the incoporation of the ligand L1Ig6, which binds the integrin $\alpha_v\beta_3$, was used to amplify the angiogenic response. More recently, Pompe et al.[91] studied how the binding of synthetic substrates to fibronectin influenced endothelial cell behavior *in vitro*. Their results indicate that fine tuning the ligand density of fibronectin allowed for the controlled formation of vascular-like structures. Overall, these studies highlight how modifying the matrix components of a construct may also useful for engineering microvessels into constructs before implantation.

Finally, the formation of endothelial cell-lined tubes represents only a very early step in the assembly of mature capillaries, arterioles, and venules. These vessels are surrounded by pericytes and smooth muscle and it is now clear that the presence of these abluminal cells is critical for complete vessel maturation. To address this need, Shepherd et al.[92] have developed a method for culturing microvessel fragments *in vitro*. Subsequent to their isolation from rat fat and seeding within collagen gels *in vitro*, Shepherd et al.[92] reported that these vessel fragments undergo angiogenic sprouting within 5 days. Upon implantation into SCID mice, vessel fragments inosculated with the host vasculature and smooth muscle staining revealed that, following the appearance of flow in the constructs, these networks contained mature arterioles and venules with smooth muscle derived from the construct. Ultimately, this study highlights the importance of vessel maturation for tissue engineering and the potential for using such approaches for tissue engineering microvessels *in vitro*.

## REFERENCES


1. Simons M, Bonow R, Chronos NA, Cohen DJ, Giordano FJ, Hammond K, Laham RJ, Li W, Pike M, Sellke FW, Stegmann TJ, Udelson JE, and Rosengart TK (2000) Clinical trials in coronary angiogenesis: issues, problems, consensus. Circulation. 102:e73-e86.
2. Freedman, SB, and JM Isner (2002) Therapeutic angiogenesis for coronary artery disease. Ann Intern Med. 136:54-71.
3. Lee, MS, and RR Makkar (2004) Stem-cell transplantation in myocardial infarction: A status report. Ann Intern Med. 140:729-737.
4. Chen, RR, and DJ Mooney (2003) Polymeric growth factor delivery strategies for tissue engineering. Pharmaceutical Res 20:1103-1112.
5. Tabata, Y, and Y Ikada (1999) Vascularization effect of basic fibroblast growth factor released from gelatin hydrogels with different biodegradabilities. Biomaterials. 20:2169-2175.
6. Thompson, JA, KD Anderson, JM Dipietro, JA Zweibel, M Zametta, WF Anderson, and T Maciag (1988) Site-directed neovessel formation *in vivo*. Science. 241:1349-1352.
7. Tabata, Y, S Hijikata, and Y Ikada (1994) Enhanced vascularization and tissue granulation by basic fibroblast growth factor impregnated in gelatin hydrogels. J Control Rel. 31: 189-199.
8. Drury, JL, and DJ Mooney (2003) Hydrogels for tissue engineering: scaffold design variables and applications. Biomaterials. 24:4337-4351.
9. Lee, KY, MC Peters, KW Anderson, and DJ Mooney (2000) Controlled growth factor release from synthetic extracellular matrices. Nature. 408:998-1000.
10. Lee, KY, MC Peters, and DJ Mooney (2003) Comparison of vascular endothelial growth factor and basic fibroblast growth factor on angiogenesis in SCID mice. J Control Rel. 87:49-56.
11. Murphy, WL, MC Peters, DH Kohn, and DJ Mooney (2000) Sustained release of vascular endothelial growth factor from mineralized poly(lactide-co-glycolide) scaffolds for tissue engineering. Biomaterials. 21:2521-2527.

12. Peters, MC, PJ Polverini, and DJ Mooney (2002) Engineering vascular networks in porous polymer matrices. J Biomed Mater res 60:668-678.
13. Smith, MK, MC Peters, TP Richardson, JC Garbern, and DJ Mooney (2004) Locally enhanced angiogenesis promotes transplanted cell survival. Tissue Engineering. 10: 63-71.
14. Richardson, TP, MC Peters, AB Ennett, and DJ Mooney (2001) Polymeric system for dual growth factor delivery. Nature Biotechnology. 19:1029-1034.
15. Langer, R. (1981) Polymers for the sustained release of macromolecules: Their use in a single-step method of immunization. Meth Enz. 73:57-74.
16. Fajardo, LF, J Kowalski, HH Kwan, SD Prionas, and AC Allison (1988) The disc angiogenesis system. Laboratory Investigation. 6:718-724.
17. Lopez, JJ, ER Edelman, A Stamler, MG Hibberd, P Prasad, KA Thomas, J DiSalvo, RP Caputo, JP Carozza, PS Douglas, FW Sellke, and M Simons (1998) Angiogenic potential of perivascularly delivered aFGF in a porcine model of chronic ischemia. Am J Physiol: Heart Circ Physiol. 274:H930-H936.
18. Tabata, Y, Y Matsui, and Y Ikada (1998) Growth factor release from amylopectin hydrogel based on copper coordination. J Control Rel. 56:135-148.
19. Tabata, Y, M Miyao, M Ozeki, et al. (2000) Controlled release of vascular endothelial growth factor released by use of collagen hydrogels. J Biomat Sci. 11:915-930.
20. Kipshidze, N, C Paramjoth, and M Keelan (1999) Fibrin meshwork as a carrier for delivery of angiogenic growth factors in patients with ischemic limb. Mayo Clin Proc. 74:847-848.
21. Naughton, GK, JN Mansbridge, G Gentzkow. (1992) A metabolically active human dermal replacement for the treatment of diabetic foot ulcers. Artif Organs. 21:1203-1210.
22. Pinney, E, K Liu, B Sheeman, and J Mansbridge (2000) Human three-dimensional fibroblast cultures express angiogenic activity. J Cell Physiol. 183:74-82.
23. Kellar, RS, LK Landeen, BR Shepherd, GK Naughton, A Radcliffe, and SK Williams (2001) Scaffold based three-dimensional human fibroblast culture provides a structural matrix that supports angiogenesis in infarcted heart tissue. Circulation. 104:2063-2068.
24. Lu, Y, J Shansky, M Del Tatto, P Ferland, X Wang, and H Vandenburgh (2001) Recombinant vascular endothelial growth factor secreted from tissue-engineered bioartificial muscles promotes localized angiogenesis. Circulation. 104:594-599.
25. Lu, Y, J Shansky, M Del Tatto, P Ferland, S McGuire, J Marszalkowski, M Maish, R Hopkins, X Wang, P Kosnik, M Nackman, A Lee, B Creswick, and H Vandenburgh. (2002) Therapeutic potential of implanted tissue-engineered bioartificial muscles delivering recombinant proteins to sheep heart. Ann NY Acad Sci. 961:78-82.
26. Hosaka, A, H Koyama, T Kushibiki, Y Tabata, N Nishiyama, T Miyata, H Shigematsu, T takato, and H Nagawa (2004) Gelatin hydrogel microspheres enable pinpoint delivery of basic fibroblast growth factor for the development of functional collateral vessels. Circulation. 110:3322-3328.
27. Blomley MJ, Cooke JC, Unger E, et al. (2001) Microbubble contrast agents: a new era in ultrasound. BMJ 322:1222-1225.
28. Mulvaugh SL, DeMaria AN, Feinstein SB, et al. (2000) Contrast echocardiography: current and future applications. J Am Soc Echocardiography 13:331-342.
29. Mayer S, Grayburn PA. (2001) Myocardial contrast agents: recent advances and future directions. Progress in Cardiovascular Diseases 44:33-44.
30. Kaul S. (2001) Myocardial Contrast echocardiography: basic principles. Progress in Cardiovascular Diseases 44:1-11.

31. Shohet RV, Chen S, Zhou Y-T, et al. (2000) Echocardiographic destruction of albumin microbubbles directs gene delivery to the myocardium. Circulation 101:2554-2556.
32. Mukherjee D, Wong J, Griffin B, et al. (2000) Ten-fold augmentation of endothelial uptake of vascular endothelial growth factor with ultrasound after systemic administration. J Am Coll Cardiol 35:1678-1686.
33. Price RJ, Skyba DM, Kaul S, et al. (1998) Delivery of colloidal particles and red blood cells to tissue through microvessel ruptures created by targeted microbubble destruction with ultrasound. Circulation 98:1264-1267.
34. Song J, Chappell JC, Qi M, et al. (2002a) Influence of Injection Site, Microvascular Pressure, and Ultrasound Variables on Microbubble Mediated Delivery of Microspheres to Muscle. J Am Coll Cardiol 39:726-731.
35. Kondo, I, K Ohmori, A Oshita, H takeuchi, S Fuke, K Shinomiya, T Noma, T Namba, M Kohno (2004) Treatment of acute myocardial infarction by hepatocyte growth factor gene transfer: The first demonstration of myocardial transfer of a "functional" gene using ultrasonic microbubble destruction. J Am Coll Cardiol. 44:644-653.
36. Song J, Qi M, Kaul S, et al. (2002b) Stimulation of Arteriogenesis in Skeletal Muscle by Microbubble Destruction with Ultrasound. Circulation 106:1550-1555.
37. Song J, PS Cottler, AL Klibanov, S Kaul, and RJ Price (2004) Microvascular remodeling and accelerated hyperemia blood flow restoration in arterially occluded skeletal muscle exposed to ultrasonic microbubble destruction. Am J Physiol: Heart Circ Physiol. 287:H2754-H2761.
38. Villanueva FS, Jankowski RJ, Klibanov S, et al. (1998) Microbubbles targeted to intercellular adhesion molecule-1 bind to activated coronary artery endothelial cells. Circulation 98:1-5.
39. Lindner JR, Coggins MP, Kaul S, et al. (2000a) Microbubble persistence in the microcirculation during ischemia/reperfusion and inflammation is caused by integrin- and complement-mediated adherence to activated leukocytes. Circulation 101:668-675.
40. Lindner JR, Song J, Xu F, et al. (2000b) Noninvasive ultrasound imaging of inflammation using microbubbles targeted to activated leukocytes. Circulation 102:2745-2750.
41. Lindner JR, Dayton PA, Coggins MP, et al. (2000c) Noninvasive imaging of inflammation by ultrasound detection of phagocytosed microbubbles. Circulation 102:531-538.
42. Dayton PA, Chomas JE, Lum AF, et al. (2001) Optical and acoustical dynamics of microbubble contrast agents inside neutrophils. Biophysical J 80:1547-1556.
43. Dalecki D, CH Raeman, SZ Child, et al. (1997a) The influence of contrast agents on hemorrhage produced by lithotripter fields. Ultra Med Biol 23:1435-1439.
44. Dalecki D, CH Raeman, SZ Child, et al. (1997b) Remnants of Albunex® nucleate acoustic cavitation. Ultra Med Biol 23:1405-1412.
45. Gamarra F, F Spelsberg, M Dellian, et al. (1993) Complete tumor remission after therapy with extra-corporeally applied high-energy shock waves. Intl J Cancer 55:153-156.
46. Miller DL, Gies RA. (1998) Gas-body-based contrast agent enhances vascular bioeffects of 1.09 MHz ultrasound on mouse intestine. Ultrasound Med Biol 24:1201-1208.
47. Skyba DM, Price RJ, Linka AZ, et al. (1998) Direct *in vivo* visualization of intravascular destruction of microbubbles by ultrasound and its local effects on tissue. Circulation 98:290-293.
48. Teupe C, S Richter, B Fissithaler, V Randriamboavonjy, C Ihling, I Fleming, R Busse, AM Zeiher, and S Dimmeler (2002) Vascular gene transfer of phosphomimetic

endothelial nitric oxide synthase (S1177D) using ultrasound-enhanced destruction of plasmid-loaded microbubbles improves vasoreactivity. Circulation. 105:1104-1109.

49. Tamiyama, Y, K Tachibana, K Hiraoka, T Namba, K Yamasaki, N Hashiya, M Aoki, T Ogihara, K Yasufumi, and R Morishita (2002) Local delivery of plasmid DNA into rat carotid artery using ultrasound. Circulation. 105:1233-1239.

50. Malekan R, Reynolds C, Narula N, et al. (1998) Angiogenesis in transmyocardial revascularization: a nonspecific response to injury. Circulation 98 (Suppl):II62-II65.

51. Victor C, Jin-qiang K, McGinn A, et al. (1999) Angiogenic response induced by mechanical transmyocardial revascularization. J Thorac Cardiovasc Surg 118:849-856.

52. Horvath KA, Chiu E, Maun D, et al. (1999) Up-regulation of vascular endothelial growth factor mRNA and angiogenesis after transmyocardial laser revascularization. Ann Thorac Surg 68: 825-829.

53. Hughes GC, Lowe JE, Kypson AP, et al. (1998) Neovascularization after transmyocardial revascularization in a model of chronic ischemia. Ann Thorac Surg 68:2029-2036.

54. Yamamoto N, Kohmoto T, Gu A, et al. (1998) Angiogenesis is enhanced in ischemic canine myocardium by transmyocardial laser revascularization. J Am Coll Cardiol 31:1426-1433.

55. Domkowski PW, Biswas SS, Steenbergen C, et al. (2001) Histological evidence of angiogenesis 9 months after transmyocardial laser revascularization. Circulation 103:469-471.

56. Price, RJ, and S Kaul (2002) Contrast ultrasound targeted drug and gene delivery: An update on a new therapeutic modality. J Cardiovasc Pharmacol Therapeut 7:171-180.

57. Cleland, JL, ET Duenas, A Park, A Daugherty, J Kahn, J Kowalski, and A Cuthbertson (2001) Development of poly-(D,L-lactide-coglycolide) microsphere formulations containing recombinant human vascular endothelial growth factor to promote local angiogenesis. J Control Rel 72:13-24.

58. Downs, EC, NE Robertson, TL Riss, and ML Plunkett (1992) Calcium alginate beads as a slow-release system for delivering angiogenic molecules *in vivo* and *in vitro*. J Cell Physiol. 152:422-429.

59. Gu, F, B Amsden, and R Neufeld (2004) Sustained delivery of vascular endothelial growth factor with alginate beads. J Control Rel. 96:463-472.

60. Sakakibara, Y, K Nishimura, K Tambara, M Yamamoto, F Lu, Y Tabata, and M Komeda (2002) Prevascularization with gelatin microspheres containing basic fibroblast growth factor enhances the benefits of cardiomyocyte transplantation. J Thoracic Cardiovasc Surgery. 124:50-56.

61. Kasahara, H, E Tanaka, N Fukuyama, E sato, H Sakamto, Y Tabata, K Ando, H Iseki, Y Shinozaki, K Kimura, E Kuwabara, S Koide, H Nakazawa, and H. Mori (2003) Biodegradable gelatin hydrigel potentiates the angiogenic effect of fibroblast growth factor 4 plasmid in rabbit hindlimb ischemia. J Am Coll Cardiol. 41:1056-1062.

62. Sieminski AL, and KL Gooch (2000) Biomaterial-microvascular interactions. Biomaterials. 21:2233-2241.

63. Padera, RF, and CK Colton (1996) Time course of membrane microarchitecture-driven neovascularization. Biomaterials. 17:277-284.

64. Brauker JH et al. (1995) Neovascularization of synthetic membranes directed by membrane microarchitecture. J Biomed Mater Res. 29:517-1524.

65. Sharkawy AA et al. (1997) Engineering the tissue which encapsulates subcutaneous implants. J Biomed Mater Res. 37:401-412.

66. Hodde, J (2002) Naturally occurring scaffolds for soft tissue repair and regeneration. Tissue Engineering. 8:295-307.

67. Kropp BP, SF Badylak, and KB Thor (1995) Regenerative bladder augmentation: a review of the initial preclinical studies with porcine small intestine submucosa. In: Zedric, S., ed. Muscle, Matrix, and Bladder Function. New York. Plenum press, pp. 229-235.
68. Wainwright, DJ (1995) Use of an acellular allograft dermal matrix (AlloDerm) in the management of full-thickness burns. Burns. 21:243-248.
69. Curtis, RJ, JC Delee, and DJ Drez (1985) Reconstruction of the anterior cruciate ligament with freeze dried fascia lata allografts in dogs. Am J Sports Med. 13:408-414.
70. Reddy, BP, DJ Barrieras, G Wilson, et al. (2000) Regeneration of functional bladder substitutes using large segment acellular matrix allografts in a porcine model. J Urol. 164:936-941.
71. Anderson, CR, AM Ponce, and RJ Price (2004) Immunohistochemical identification of an extracellular matrix scaffold that microguides capillary sprouting *in vivo*. J Histochem Cytochem. 52:1063-1072.
72. Thompson, JA et al. (1989) Heparin-binding growth factor-1 induces the formation of organoid neovascular structures *in vivo*. Proc Nat Acad Sci. 86:7928-7932.
73. Pandit, AS, DS Feldman, J Caulfield, and A Thompson (1998) Stimulation of angiogenesis by FGF-1 delivered through a modified fibrin scaffold. Growth Factors. 15:113-123.
74. Fournier, N, and CJ Doillon (1996) Biological molecule-impregnated polyester: an in-vivo angiogenesis study. Biomaterials. 17:1659-1665.
75. Masuda, S et al. (1997) Vascular endothelial growth factor enhances vascularization in microporous small caliber polyurethane grafts. ASAIO J. 43:M530-534.
76. Kidd, KR, RB Nagle, and SK Williams (2002) Angiogenesis and neovascularization associated with extracellular matrix-modified porous implants. J Biomed Mater Res. 59:366-377.
77. Kidd, KR, and SK Williams (2004) Laminin-5-enriched extracellualr matrix accelerates angiogenesis and neovascularization in association with ePTFE. J Biomed Mater Res. 69A:294-304.
78. Nör, JE, JB Christensen, DJ Mooney, and PL Polverini (1999) Vascular endothelial growth factor (VEGF)-mediated angiogenesis is associated with enhanced endothelial cell survival and induction of Bcl-2 expression. Am J Pathol. 154:375-384.
79. Yang, J, U Nagavarapu, K Relioma, MD Sjaastad, WC Moss, A Passaniti, and GS Heron (2001) Telomerized human microvasculature is functionalized *in vivo*. Nature Biotechnology. 19:219-224.
80. Enis, DR, BR Shepherd, Y Wang, A Qasim, CM Shanahan, PL Weissberg, M Kashgarian, JS Pober, and JS Schechner (2005) Induction, differentiation, and remodeling of blood vessels after transplantation of Bcl-2-transduced endothelial cells. Proc Nat Acad Sci. 102:425-430.
81. Nör, JE, MC Peters, JB Christensen, MM Sutorik, S Linn, MK Khan, CL Addison, DJ Mooney, and PJ Polverini (2001) Engineering and characterization of functional human microvessels in immunodeficient mice. Lab.Invest. 81:453-463.
82. Levenburg, S, JS Golub, M Amit, J Itskovitz-Eldor, and R Langer (2002) Endothelial cells derived from human embryonic stem cells. Proc Nat Acad Sci. 99:4391-4396.
83. Wang, H-J, J Pieper, R Schotel, CA Van Blitterswijk, and EN Lamme (2004) Stimulation of skin repair is dependent on fibroblast source and presence of extracellular matrix. Tissue Engineering. 10:1054-1064.
84. Kawai, K, S Suzuki, Y Tabata, Y ikada, and Y Nishimura (2000) Accelerated tissue regeneration through incorporation of basic fibroblast growth factor-impregnated gelatin microspheres into artificial dermis. Biomaterials. 21:489-499.

85. Lee, H, RA Cusick, F Browne, TH Kim, PX Ma, H Utsunomiya, R Langer, and JP Vacanti (2002) Local delivery of basic fibroblast growth factor increases both angiogenesis and engraftment of hepatocytes in tissue-engineered polymer devices. Transplantation. 73:1589-1593.
86. Gardner-Thorpe, J, TV Grikscheitt, H Ito, A perez, SW Ashley, JP Vacanti, and EE Whang (2003) Angiogenesis in tissue-engineered small intestine. Tissue Engineering. 9:1255-1261.
87. Black, AF, F Berthod, N L'Heureux, L Germain, and FA Auger (1998) *In vitro* reconstruction of a human capillary-like network in a tissue-engineered skin equivalent. FASEB J. 12:1331-1340.
88. Hudon, V, F Berthod, AF Black, O Damour, L Germain, and FA Auger (2003) A tissue-engineered endothelialized dermis to study the modulation of angiogenic and angiostatic moleculaes on capillary-like tube formation *in vitro*. British J Dermatol. 148:1094-1104.
89. Frerich, B, N Lindemann, J Kurtz-Hoffman, and K Oertel (2001) *In vitro* model of a vascular stroma for the engineering of vascularized tissues. Int J Oral Maxillofacial Surg. 30:414-420.
90. Hall, H, T Baechi, and JA Hubbell (2001) molecular properties of fibrin-based matrices for promotion of angiogenesis *in vitro*. Microvasc Res. 62:315-326.
91. Pompe, T, M Markowski, and C Werner (2004) Modulated fibronectin anchorage at polymer substrates controls angiogenesis. Tissue Engineering. 10:841-848.
92. Shepherd, BR, HYS Chen, CM Smth, G Gruionu, SK Williams, and JB Hoying (2004) Rapid perfusion and network remodeling in a microvascular construct after implantation. Arterioscler Thromb Vasc Biol. 24:898-904.

Chapter 8

# LYMPHANGIOGENESIS: RECAPITULATION OF ANGIOGENESIS IN HEALTH AND DISEASE

William S. Shin, Stanley G. Rockson
*Stanford Center for Lymphatic and Venous Disorders, Division of Cardiovascular Medicine, Stanford University School of Medicine*

## 1. INTRODUCTION

The lymphatic vascular system is designed to maintain tissue fluid equilibrium by providing a vital route for protein and fluid transport, and plays a complementary role with the blood vasculature in tissue perfusion and fluid reabsorption. The lymphatics also serve as conduits for intestinal lipid absorption and for transport of lymphocytes and antigen-presenting dendritic cells to regional lymph nodes. The lymphatic system is involved in a variety of disease processes. Hypoplasia or dysfunction of the lymphatic vasculature can lead to a pathologic condition termed lymphedema, whereas hyperplasia or abnormal development of these vessels is associated with lymphangiomas and lymphangiosarcomas. Moreover, a role for lymphatic vessels is invoked in the process of nodal and systemic metastasis of cancer cells. Despite the critical role of lymphatics in mediating tissue fluid homeostasis, intestinal lipid absorption, and the immune response, our understanding of the lymphatic system has, until recently, been limited by the absence of specific molecular markers for identifying lymphatics and relatively poor insight into the molecular participants in lymphatic functional regulation. Recently, the discovery and characterization of lymphatic-specific growth factors, receptors, and transcriptional regulators have afforded insights into the mechanisms of lymphatic development and disease. Current studies of genetic animal models have established a new molecular model of embryonic lymphatic vascular development, have



*R. Forough (ed.), New Frontiers in Angiogenesis,* 159–202.

identified molecular pathways leading to human diseases associated with lymphedema; and have provided evidence that lymphatic vessel activation may be important in tumor dissemination. These advances offer promise for the development of targeted molecular therapies for the treatment of lymphatic disease.

## 2. THE LYMPHATIC SYSTEM: STRUCTURE AND FUNCTION

The lymphatic system is a component of both the circulatory and the immune systems[1]. In addition to mediating tissue fluid homeostasis by draining protein-rich lymph from tissues and transporting it to the blood vascular system for recirculation, the lymphatic system also serves as a conduit for the transport of lymphocytes and antigen-presenting dendritic cells to regional lymph nodes[2]. The lymphatic and the blood vascular systems operate in parallel and share anatomic features, but they also exhibit distinct structural and functional characteristics. Blood capillary endothelial cells are surrounded by a continuous basement membrane and by smooth muscle cells/pericytes, while lymphatic capillaries are blind-ended endothelial tubes lacking pericytes and a continuous basal lamina, but containing large interendothelial valve-like openings and anchoring filaments that connect the vessels to the extracellular matrix[3-9]. Whereas the blood vascular system is a closed circulation in which blood has egress from, and returns to, the same system, the lymphatic vasculature comprises an open-ended network through which lymph is drained from the interstitial space of tissues and is transported from initial lymphatic capillaries to larger collecting vessels and, finally, to the inferior vena cava for recirculation[2].

The lymphatic vasculature is comprised of initial lymphatics and their precollectors, which coalesce into lymphatic ducts that drain into lymph nodes[1,10]. Initial lymphatics exist in the skin as blind-end sinuses[11,12] which form a superficial subpapillary plexus of interconnected sinuses. This plexus consists of single layers of gracile lymphatic endothelial cells[13,14] which rest on a discontinuous basement membrane. The lymphatic basement membrane is composed of type IV collagen and, unlike the vascular capillary basement membrane, contains no heparan sulfate, proteoglycan, or fibronectin[15]. The initial lymphatics are significantly larger in diameter than arteriovenous capillaries (8 um), ranging in size from 10 to 60 um[14,16]. Interendothelial openings permit extracellular fluid, macromolecules and cells to drain directly into the lumina of the initial lymphatics through the porous basement membrane[17-19]; there are no tight junctions between the cells. Estimates of the pore size, based on measurements of intercellular junctional distances, vary from 15 nm to several micrometers[20,21]. Interendothelial junctions form an interdigitated, overlapping structure that provide a one-way valve system for fluid movement[18]. Such endothelial clefts open to dimensions of up to several micrometers, allowing macromolecules, colloids, cells, and cellular debris to pass depending on the extent of distension[15,19,22,23]. Interendothelial junctions open by in-plane stretching of the lymphatic endothelium as a consequence of fluid inflow from the interstitium or edema. Reflux of lymphatic fluid into the interstitium is prevented, in theory, by resealing of the endothelial clefts.

The initial lymphatics are connected in a hexagonal pattern , through lymphatic precollectors, with the deeper lymphatics in the dermis[1]. Lymph is subsequently transported centrally through collecting ducts and then to the lymph nodes. Like the initial lymphatics, the superficial precollectors have no detectable vasomotor activity, an observation consistent with ultrastructural studies that depict a fine endothelial lining without smooth muscle[4,19,24]. The collecting ducts, into which the precollectors coalesce, have thick walls containing a thin layer of smooth muscle that is separated from the vessel lumen by a monolayer of endothelial cells[14,25]. All collecting lymphatics contain bicuspid or unicuspid valves which prevent the reflux of lymph[10,14,23,26,27].

Under normal physiologic conditions, lymphatic capillaries predominantly remain collapsed[2]. However, if interstitial pressure increases (*e.g.* due to fluid efflux from hyperpermeable blood vessels), the anchoring filaments exert tension on the lymphatic endothelial cells to pull open their overlapping cell junctions. Thus, these filaments are thought to maintain the patency of lymphatic vessels during increased tissue inflammation[9]. There is no central pump to aid in the unidirectional flow of fluid in the lymphatic vessel network. Lymph is driven in the tissues primarily through the compression of lymphatic vessels by adjacent skeletal muscles. Contractility of the collecting lymphatic vessel wall in response to stretch also appears to contribute to lymphatic vessel function[28]. The valves of the lymphatics further assist in the unidirectional flow of lymph.

Lymphatics are found in most vascularized tissues and organs throughout the body, with the exception of the brain and the retina[1,2]. In the central nervous system, cerebrospinal fluid fulfills the normal role of lymph. Although the cornea is normally devoid of both blood and lymphatic vessels, significant lymphangiogenesis, and angiogenesis, can occur after surgical manipulation and in several corneal diseases[29]. Lymphatic tissue is prevalent in organs that are exposed to direct contact with the external environment, such as the skin, lungs, and gastrointestinal tract; this distribution presumptively reflects the role that the lymphatics enjoy in the protection against potential pathogens. In the gastrointestinal tract, the lymphatic system mediates the absorption of fat from the intestine and transports the lipids, in the form of chyle, to the liver. In addition, as previously described, the lymphatic system also transports fluid excesses, along with cellular debris and metabolic waste products from peripheral sites, back to the systemic circulation.

In addition to the vascular structures, the lymphatic system consists of lymphoid cells and organized lymphoid tissues, including the lymph nodes, tonsils, spleen, thymus, Peyer's patches in the intestine, and lymphoid tissue in the lungs, liver, and parts of the bone marrow[1,30]. These lymphoid organs play an integral role in the immune response. Foreign substances (antigens) are concentrated by the dendritic cells and presented to lymphocytes in these specialized lymphoid tissues, which leads to a cascade of steps that produces the orchestrated immune responses[9]. It is important to recognize that lymphocytes circulate between the lymphatic and blood vasculatures. Lymphocytes in circulating lymph enter lymph nodes via the afferent lymphatic vessels, while lymphocytes present in blood may access the lymph node through the wall of special postcapillary venules. Lymphocytes, along with lymph, are subsequently recirculated to the blood circulation through the efferent lymphatic vessels and the thoracic duct.

The trafficking of lymphocytes, antigen presenting cells, and dendritic/Langerhans cells from peripheral tissues to the lymphatic vessels is mediated by adhesion molecules including integrins, selectins and their ligands[9,31-34], and also by chemokine signaling. For instance, dendritic cell activation is accompanied by the upregulation of the cytokine receptor CCR7 and, subsequently, by dendritic cell sensitization to secondary lymphoid chemokine (SLC) that is constitutively produced by skin lymphatic endothelial cells[35,36]. Interestingly, certain human breast cancer cell lines have been demonstrated to express the CCR7 receptor[37]. Furthermore, in an experimental murine model of melanoma, CCR7 expression was shown to enhance lymphatic metastasis of melanoma cells as compared to control tumor cells, and neutralizing anti-SLC antibodies were capable of blocking this effect[38]. It is conceivable, then, that tumor cells may share similar trafficking pathways with lymphocytes and antigen presenting cells in accessing the lymphatic vessels.

## 3. DEVELOPMENT OF THE LYMPHATIC SYSTEM

The development of the blood vascular system is well described and occurs via two processes, vasculogenesis and angiogenesis[9]. Vasculogenesis involves the *de novo* differentiation of endothelial cells (ECs) from mesoderm-derived precursor cells, termed hemangioblasts[39].

Hemangioblasts differentiate to form angioblasts that cluster and reorganize to form capillary-like tubes. Once this primary vascular plexus is formed, new capillaries form by sprouting from pre-existing vessels in a process called angiogenesis[40]. The newly formed vasculature is further remodeled into a tree-like hierarchy of vessels containing vessels of different sizes. Excess branches are pruned, some vessels fuse to form larger ones and others regress. Endothelial cells also begin to differentiate into arterial or venous types in the primary capillary plexus[41].

In contrast to the blood vascular system, the lymphatic system has historically been subject to only limited scientific attention, largely due to its indistinctive morphology and lack of identifying lymphatic-specific markers[2]. Centuries ago, Gasparo Aselli first identifed lymphatic vessels as "milky veins" in the dog mesentery[42]. Despite the subsequent identification of the collecting lymphatics and thoracic duct, the mechanisms responsible for lymphatic vasculogenesis remained unclear. In 1902, Sabin suggested that initial lymph sacs emerge by budding from embryonic veins, and that these primitive lymphatics then spread out throughout the body to form lymphatic networks[43,44]. An alternative model proposed that the initial lymph sacs arise from mesenchymal precursor cells, and that the connection to the venous system is formed later in development[45]. It now appears that both of the proposed mechanisms may contribute to the formation of the lymphatic system.

Several lymphatic-specific markers, namely, lymphatic vessel endothelial hyaluronan receptor (LYVE-1), vascular endothelial growth factor receptor-3 (VEGFR-3) and the transcription factor Prox-1, are expressed in the endothelial cells forming the budding lymph sacs in mouse embryos, supporting Sabin's theory of lymphatic development[9,46-48]. Studies of Prox1-deficient mice have revealed that these mice are unable to develop a lymphatic vascular system, and that Prox1 is required for endothelial cells in the embryonic cardinal veins to migrate and form the initial lymphatic vessels during early embryogenesis[2,48,49]. Recent studies of avian development indicate that mesodermal lymphangioblasts may independently contribute to the formation of the lymphatic vascular system in the early wing buds, limb buds, and chorioallantoic membrane of birds[50-53], supporting the alternative model of lymphatic development.

Prox1 is a homeodomain protein that has been identified in humans, mice, chicken, newt, frog, and zebra fish[2]. Prox1 amino acid sequences are highly conserved across these specie[54-56]. The Prox1 transcript is detectable in the developing liver, pancreas, nervous system, and heart[54], where it serves as a master regulator of cell fate decisions that lead to the establishment of different cell lineages during embryogenesis. Detailed analysis of Prox-1 null mice has led to the formulation of a molecular model for early lymphatic vascular development[57], in which venous endothelial cells become competent to a lymphatic-inducing signal. At mouse embryonic day 8.5 (E8.5), all cardinal vein endothelial cells display lymphatic competence in expressing LYVE-1 and VEGFR-3, both of which are later expressed specifically by lymphatic endothelium. At E9.5–E10.5, an unknown inductive signal derived from the surrounding mesenchyme stimulates a subset of endothelial cells situated on one side of the cardinal vein to express Prox1[48,49]. At this stage, LYVE-1 is uniformly expressed in the cardinal vein endothelial cells. At E10.5-E11.5, these Prox-1 and Lyve-1 double positive cells bud off and migrate out in a polarized manner[49] to form the initial rudimentary lymph sacs, from which lymphatic vessels spread to peripheral tissues of the embryo[48]. Thus all venular endothelial cells may initially be bipotent, but upon the asymmetric expression of at least Prox-1 in a restricted population, these cells become committed to lymphatic differentiation[9,49].

As the lymphatic endothelial precursor cells bud in a polarized manner, they alter their gene expression profiles to adopt a lymphatic phenotype. This is accompanied by expression of additional lymphatic markers such as SLC and neuropilin-2, and by progressivedown-regulation of blood vascular endothelial markers, including CD34, laminin, and type IV collagen[2,49]. VEGFR-3 expression is maintained at a high level in the budding lymphatic endothelial cells and the simultaneous expression of LYVE-1, VEGFR-3, Prox-1, and SLC may indicate irreversible commitment to the lymphatic endothelial cell lineage[9,49]. From E12.5 on, Prox1-positive, lymphatically-specified endothelial cells further differentiate and sprout throughout the embryo, while remaining Prox1-negative venous endothelial cells lose LYVE-1 and VEGFR-3 expression to adopt a blood vascular endothelial cell phenotype.

The observation that the initial budding of immature lymphatic endothelial cells from the cardinal vein of Prox1-deficient mice remains detectable, but that their lineage-specific lymphatic specification does not occur[48,49] has suggested that Prox1 may specify lymphatic cell fate by directly reprogramming the transcriptome of embryonic venous endothelial cells[2]. Indeed, the ectopic expression of Prox1 in differentiated blood vascular endothelial cells has been shown to be sufficient to reprogram these cells to adopt a lymphatic phenotype[2]. This reprogramming is associated with repression of blood vascular endothelial cell-specific genes including E-selectin, laminin, and neuropilin-1, and with induction of several lymphatic-specific genes including podoplanin and VEGFR-3[58,59].

## 4. PHENOTYPIC CHARACTERIZATION OF THE LYMPHATIC ENDOTHELIAL CELL

The recent establishment of defined cultures of human lymphatic and blood vascular endothelial cells[60-63] has permitted the comparative analysis of their specific transcriptomes[2]. Approximately 98% of genes that have been investigated by microarray analyses appear to be expressed at comparable levels in the two endothelial cell types[58,63], corroborating their close genetic relationship. However, human lymphatic and blood vascular endothelial cells do display striking differences in the expression of pro-inflammatory chemokines and cytokines (monocyte chemotactic protein-1, interleukin-6 and interleukin 8), their receptors (UFO/axl, CXCR4, IL-4R), and genes involved in cytoskeletal and cell-cell interactions (integrins and proteins associated with cadherin junctions)[9,58,63]. Integrin alpha5 is primarily expressed in blood vascular endothelial cells, while integrin alpha9 appears to be lymphatic endothelial cell-specific[58]. Notably, mice lacking integrin alpha9beta1 develop respiratory failure due to the accumulation of a milky pleural effusion, presumably lymphatic, and die soon after birth[64]. This phenotype may relate to a co-operation between alpha9beta1 and VEGFR-3 signaling[9,65].

## 5. MOLECULAR CONTROL OF LYMPHANGIOGENESIS

Both lymphangiogenesis and angiogenesis are tightly regulated by intercellular signaling mechanisms, cytokines and growth factors. The angiogenic switch is thought to be caused by a shift in the net balance of positively acting angiogenic mediators and negatively acting angiogenesis inhibitors[9]. The mechanisms that contribute to this shift of balance are incompletely understood, but several factors including hypoxia and, in tumors, oncogenes and tumor suppressor genes, are known to contribute to this mechanism by up- or down-regulating pro-angiogenic growth factors and endogenous angiogenesis inhibitors[66,67]. Although the regulation of lymphangiogenesis is less well understood, the recent identification of VEGF-C/D as the primary mediators of lymphangiogenesis has provided much insight into the signaling mechanisms governing this process.

# 6. THE VEGF FAMILY OF GROWTH FACTORS: IMPLICATIONS IN ANGIOGENESIS AND LYMPHANGIOGENESIS

The members of the vascular endothelial growth factor (VEGF) family are integrally involved in the development of blood vascular and lymphatic structures[9]. Five identified growth factors, namely VEGF (or VEGF-A), VEGF-B, VEGF-C, VEGF-D and PIGF constitute this family. Viral homologues, including the orf virus VEGF (or VEGF-E), have additionally been detected. The VEGFs are secreted dimeric glycoproteins that contain characteristic, regularly spaced cysteine knot motifs. These growth factors bind to three known VEGF receptors, VEGFR-1, VEGFR-2 and VEGFR-3, with varying specificities.

VEGF has several isoforms of differing amino acid chain lengths. All VEGF isoforms bind to VEGFR-1 and VEGFR-2[68-72], with VEGFR-2 serving as the main receptor for VEGF in blood vascular endothelial cells[9]. As a mitogen, VEGF is highly specific for blood vascular endothelial cells[73-80] and is widely expressed in tissues immediately adjacent to areas of active vessel formation in vertebrate embryos[81]. VEGF expression is critical for the earliest stages of vasculogenesis, as blood islands, endothelial cells, and major vessel tubes fail to develop in VEGF knockout embryos[82,83], with death occurring at mid-gestation. Surprisingly, inactivation of a single VEGF allele in mice also results in embryonic lethality due to defective, though not abolished, vasculogenesis and angiogenesis[82,83]. This heterozygous lethal phenotype is indicative of a tight dose-dependent regulation of embryonic vessel development by VEGF[2]. VEGF is furthermore a survival factor for endothelial cells, as it has been demonstrated to induce the expression of anti-apoptotic factors *in vivo* and *in vitro*[9,84-87]. Pericyte coverage of newly formed blood vessels appears to be the critical event that determines when endothelial cells no longer require VEGF for survival *in vivo*[88]. VEGF is also well known to be a potent inducer of vascular leak[89-91].

VEGF-B has two differentially expressed isoforms, both of which serve as specific ligands for VEGFR-1[92-94]. Though mice lacking VEGF-B do not have an obvious phenotype[95], VEGF-B, through interaction with VEGFR-1, may have important effects on monocyte migration, mobilization of bone marrow derived VEGFR-1 positive stem cells, and pathological angiogenesis via recruitment of myeloid and endothelial precursur cells and inflammatory cells[96-98]. In this regard, VEGF-B exhibits functional similarity with PIGF, another member of the VEGF family that binds specifically to VEGFR-1[99,100]. PIGF, like VEGF-B, has two isoforms[99,101-103]. PIGF homodimers are chemotactic for monocytes and endothelial cells in culture[104], though PIGF alone is not capable of inducing endothelial cell proliferation or vascular permeability[100]. As with VEGF-B, PIGF deficient mice do not have an obvious phenotype[95]. Loss of PIGF, however, has been shown to impair angiogenesis and collateral growth during ischemia, inflammation, wound healing and cancer[95]. Transplantation with wild type bone marrow rescues the impaired angiogenesis in PIGF deficient mice, indicating that PIGF may contribute to blood vessel growth by mobilizing bone-marrow derived endothelial precursor cells[9,95]

## 6.1 VEGF-C and VEGF-D: The Primary Regulators of Lymphangiogenesis

VEGF-C and VEGF-D, originally cloned as ligands for VEGFR-3, were the first factors identified with the capacity of inducing the growth of new lymphatic vessels *in vivo*[105-109]. VEGF-C is produced as a preproprotein with long N- and C terminal propeptides flanking the VEGF homology domain, and a series of proteolytic cleavage steps are needed to generate a fully processed form with high affinity for both VEGFR-2 and VEGFR-3[110]. VEGF-C is expressed by numerous cell types, including mesenchymal cells in regions where lymphatic vessels sprout from embryonic veins, activated macrophages, smooth muscle cells surrounding large arteries, and skeletal muscle cells[105,111-113]. In developing mouse embryos, VEGF-C is highly expressed in the mesenterium, lung, heart and kidney[111]. In human tissues, VEGF-C has been detected in neuroendocrine cells of fenestrated blood vessels[114]. Tumor necrosis factor alpha and interleukin-1 beta upregulate VEGF-C mRNA, while dexamethasone and interleukin-1 receptor antagonists appear to inhibit this effect[115]. Interestingly, at least two reports have documented VEGF-C expression in blood vascular endothelial cells; during avian embryonic development and in Kaposi's sarcoma[112,116], suggesting possible paracrine mechanisms by which the blood vascular system may control lymphatic vessel growth and maintenance.

Both VEGF-C and VEGF-D promote, through activation of VEGFR-3, the proliferation of cultured human lymphatic endothelial cells[61]. VEGF-C/D, through VEGFR-3, are able to induce *in vivo* lymphangiogenesis as well, as seen in transgenic mice that overexpress VEGF-C/D[117,118]. Conversely, inhibition of lymphatic growth is obtained when VEGF-C/VEGF-D binding to endogenous receptors is blocked in transgenic mice expressing a soluble form of VEGFR-3[119]. VEGF-C has additionally been shown to be a highly specific lymphangiogenic factor in the mature chorioallantoic membrane[8].

VEGF-C plays an essential role during embryonic lymphangiogenesis, as indicated by the phenotype of VEGF-C deficient mice[113] in which early lymphatic vessel formation is completely abrogated. Although Prox1 expression is not inhibited in the lymphatically-specified cardinal vein endothelial cells of VEGF-C deficient mice, Prox1-positive cells are unable to bud out to form the initial lymph sacs, suggesting that lymphatic endothelial cell specification and subsequent cell migration are controlled by distinct signaling pathways[59]. The sprouting defect in VEGF-C negative, committed lymphatic endothelial cell progenitors is rescued in the presence of VEGF-C or VEGF-D, but not VEGF-A, indicating the necessity of VEGFR-3 mediated signaling for initial lymphatic vessel formation[113]. Mice heterozygous for VEGF-C develop cutaneous lymphatic hypoplasia and lymphedema, revealing a haploinsufficiency effect of VEGF-C for regular lymphatic function and development[59].

While VEGF-C appears to act primarily as a lymphangiogenic factor *in vivo* it might also, through potential interaction with VEGFR-2, play an important role in angiogenesis. VEGF-C has been shown to promote angiogenesis in rabbit ischemic hindlimb, in avascular mouse cornea, and in the early chorioallantoic membrane where lymphatic vessels have not yet developed[120,121]. The fully processed form of VEGF-C further induces vascular permeability via VEGFR-2[107,110]. Interestingly, while lymphatic vessel formation is abrogated in VEGF-C null embryos, no major malformations of the blood vascular system are detected[113], suggesting that molecular interactions between VEGF-C and its receptors during embryonic development of these two circulatory systems may be distinct. Alternatively, as the lymphatic and blood vascular systems are known to develop in a sequential manner, timing may be a key factor that influences the outcome of the receptor-ligand interactions[2].

VEGF-D is produced as a preproprotein and undergoes proteolytic processing and, like VEGF-C, the fully processed human VEGF-D binds to both VEGFR-2 and VEGFR-3[107,109,122]. VEGF-D, like VEGF-C, has been detected in human vascular smooth muscle cells. In developing mouse embryos, VEGF-D is expressed in several organs and structures including skin, heart, lung, and limb buds[114,123-126]. Mouse VEGF-D differs from its human counterpart in at least two aspects: it is a specific ligand for VEGFR-3 [123,124], and it is expressed as two isoforms differing in their C-termini[123,124]. This suggests that VEGF-D signaling via VEGFR-2 may not be significant for normal development and physiology, or that the biological function of VEGF-D varies in disparate species[9]. VEGF-D has been shown to be angiogenic *in vivo*[109,127], and mitogenic for cultured microvascular endothelial cells[109]. VEGF-D-deficient mice do not exhibit a lymphatic vascular phenotype[113].

## 6.2 Importance of VEGF Receptor Binding Specificity and Expression Patterns in Angiogenic/Lymphangiogenic Signaling

VEGF signaling is mediated via high-affinity receptor tyrosine kinases termed VEGFRs[9]. VEGFRs form a sub-family within the platelet derived growth factor receptor family, and share functional and structural similarities with members of this receptor class[128,129]. All VEGFRs have an extracellular ligand binding domain containing seven immunoglobulin homology regions, and an intracellular tyrosine kinase signaling domain. Upon ligand binding, the VEGFRs are thought to dimerize and undergo transphosphorylation. These phosphorylated tyrosine residues may control the kinase activity of the receptor, and create docking sites for cytoplasmic signaling molecules[9].

VEGFR-1 (or Flt1), which serves as a receptor for VEGF, VEGF-B and PIGF, is expressed either as a transmembrane protein or a shorter soluble form[130 131,132]. In adults, VEGFR-1 is expressed primarily in blood vascular endothelium, monocytes, and vascular smooth muscle[133]. Though the role of VEGFR-1 signaling in endothelial cells is poorly defined, targeted inactivation of VEGR-1 leads to overgrowth of endothelial-like cells secondary to increased hemangioblast commitment, with ensuing dis-organization of blood vasculature and embryonic lethality at E8.5[134,135]. VEGFR-1 signaling further mediates monocyte migration [136] and, according to recent data, recruitment of bone marrow derived VEGFR-1 positive hematopoeitic stem cells that may play major roles in inflammation and angiogenesis [96-98]. VEGFR-1 has not been implicated in lymphangiogenesis[9].

VEGFR-2 (or Flk1/KDR) is a receptor for VEGF, VEGF-C, and VEGF-D, and is expressed in blood vascular and lymphatic endothelial cells *in vivo* and *in vitro*[60,63,137]. VEGFR-2 serves as the main receptor for VEGF in endothelial cells[9]. Targeted disruption of VEGFR-2 results in the absence of blood vessels and blood cells, leading to lethality at E8.5-9.5[138,139]. In the mouse embryo, VEGFR-2 is first detected at E7.0 in hemangiogenic lateral plate mesoderm, but later becomes restricted to the blood islands [138]. In human postnatal tissues, VEGFR-2 continues to be expressed in endothelial cells and in hematopoietic stem cells[140]. Interestingly, arteries in mouse skin do not express VEGFR-2, though VEGFR-2 is expressed in the veins and collecting lymphatic vessels[141]. *In vitro*, VEGFR-2 activates or modifies numerous signal transduction molecules including phosphatidylinositol 3'-kinase, phospholipase Cgamma, Src family tyrosine kinases, Akt/protein kinase B, protein kinase C, and p38 mitogen-activated protein kinase in primary endothelial cells[142]. *In vivo*, VEGFR-2 activation leads to protease expression/basement membrane breakdown, the upregulation of integrins on angiogenic endothelium, and endothelial cell proliferation and migration[128]. VEGR-2 additionally mediates vascular permeability via Src kinase[143,144].

Though the role of VEGFR-2 in angiogenesis is well established, the potential involvement of VEGFR-2-mediated VEGF signaling in lymphangiogenesis is currently a matter of debate[2]. *In vitro*, VEGF is a potent inducer of lymphatic endothelial cell proliferation as demonstrated in cell proliferation assays[63]. Administration of adenoviral murine VEGF-A164 (one of the VEGF isoforms) to the mouse ear results in sustained *in vivo* lymphangiogenesis[145]. In contrast, similar studies employing adenoviral expression of the human VEGF-A165 isoform have not demonstrated distinct lymphangiogenic activity[146,147]. This observed disparity may be a reflection of tissue- or species-specificity of VEGF[2]. Recent studies involving transgenic mice have shown that skin-specific overexpression of murine VEGF-A164 results in enhancement of both angiogenesis and lymphangiogenesis during tissue repair and skin inflammation[137,148].

VEGFR-3 (or Flt4) was the first gene shown to be expressed specifically by lymphatic endothelial cells[46]. VEGFR-3 is a member of the fms-like tyrosine kinase family and is structurally related to VEGFR-1 and VEGFR-2[46,149]. However, unlike VEGFR-1 and VEGFR-2, VEGFR-3 does not interact with VEGF[2]. Rather, it serves as a signaling receptor for the lymphatic-specific growth factors VEGF-C and VEGF-D. Two splice variants of human VEGFR-3 have been characterized, a 4.5 kb transcript and a more prevalent 5.8 kb transcript[150]. The longer transcript is the major form detected in tissues[9].

During murine embryonic development, VEGFR-3 is expressed by angioblasts of the head mesenchyme and endothelial cells of the cardinal vein from E8.5 to E12.5[46]. At E12.5, VEGR-3 is detected both in developing venous and lymphatic endothelia, but its expression later becomes restricted primarily to the lymphatics[46,151-153]. Interestingly, VEGFR-3 has been detected in fenestrated blood vessel endothelia of adult human tissues, though this expression is restricted to lymphatic endothelia in tissues with continuous capillary endothelium[114].

Genetic disruption of VEGFR-3 in mice results in impaired hematopoiesis, defective remodeling of the primary vascular plexus, and embryonic lethality secondary to cardiovascular failure by E9.5, although early vasculogenesis and angiogenesis occur normally[47]. Because VEGFR-3 knockout mice die at E9.5 prior to development of the lymphatic system, the lymphatic phenotype of VEGFR-3 mutant mice has not been able to evaluate. In wild type mice, however, ablation of VEGFR-3 signaling at E13 (when VEGFR3 expression is largely restricted to the lymphatic endothelium) leads to aplasia of the lymphatics[46,111,119]. Mice heterozygous for a null allele at the VEGFR3 locus have normal-appearing lymphatics and are phenotypically normal[154].

VEGFR-3 appears to be the most critical of the VEGFRs in the context of lymphatic development[154]. The observation that lymphangiogenesis in mice can be inhibited via transgenic expression of a soluble VEGFR-3 that competes for ligand binding with the endogenous receptor indicates that VEGFR-3 signaling is required for lymphatic vasculature formation[119]. However, because VEGF-C and VEGF-D also bind to VEGFR-2 and because VEGFR-2 is also expressed by lymphatic endothelium[118,137], the distinct contributions of VEGFR-3 and VEGFR-2 to lymphangiogenesis have remained difficult to assess[2]. Studies of transgenic mice that overexpress VEGFC156S, a VEGFR-3-specific mutant of VEGF-C[155], have demonstrated that activation of VEGFR-3 signal transduction is sufficient to promote lymphangiogenesis *in vivo*[118]. Furthermore, *in vitro* experiments with VEGF-C156S have revealed that VEGFR-3 stimulation alone is sufficient to stimulate lymphatic endothelial cell migration and to protect these cells from apoptosis[61,155].

## 7. TIE RECEPTORS IN ANGIOGENESIS: IMPLICATIONS FOR LYMPHATIC DEVELOPMENT

Tie-receptors and the angiopoietins play important roles in the formation of the vascular system[156-158], and likely complement the VEGFs in the coordinated interplay of signaling interactions required for this process. Tie-1 and Tie-2 are expressed primarily by vascular endothelial cells and comprise the Tie class of receptor tyrosine kinases[9]. Both Tie-1 and Tie-2 have an extracellular domain consisting of two Ig homology regions and a cytoplasmic tyrosine kinase domain. Tie-1 is expressed by lymphatic endothelium during embryogenesis after E13.5, but its function in lymphatic endothelial cells is unknown[159]. Tie-2 expression pattern in lymphatic vessels is poorly characterized.

Though structurally related, Tie-1 and Tie-2 play distinct roles in the formation of blood vessels[160]. Tie-1 appears to be required for vascular endothelial cell survival during embryonic development. Mouse embryos deficient in Tie-1 fail to establish structural integrity of vascular endothelial cells, resulting in edema, localized microvascular hemorrhage, and death between days 13.5 and birth[160,161]. In contrast, Tie-2 appears to be important in vascular network formation and angiogenesis in endothelial cells. Deletion of Tie-2 results in insufficient expansion and maintenance of the primary capillary plexus, and consequent embryonic death between E9.5 and E10.5[160,162].

While Tie-1 does not have an identified ligand[9], the Tie-2 receptor is bound by four known angiopoietins which mediate vessel stabilization signals[163-166]. Tie-2 is activated by Ang-1 and Ang-4, while Ang-2 and Ang-3 likely function as specific antagonists for Tie-2 and inhibit its signaling at least in some settings. It appears that Ang1 acts to stabilize mature blood vessels. The phenotype of Ang1 deficient mice has suggested a role for the Ang1-Tie2 ligand-receptor system in maintaining the communication between ECs and the surrounding mesenchyme[9]. Ang-1 deletion results in angiogenic defects similar to those seen in mice lacking Tie-2, including the absence of perivascular cells[167]. Complementing these observations, adenovirally mediated Ang-1 adminsitration has been shown to protect adult vasculature against plasma leakage[168].

Ang2, which appears to serve as an antagonist for Ang1, is expressed primarily by smooth muscle cells of large arteries, veins, and venules, with enhanced expression seen at sites of active vessel remodeling such as the ovary, placenta, and tumors[164,169-172]. This expression pattern, along with the phenotypes of Ang-1 and Tie-2 knockout mice, have suggested a model for angiogenic vessel remodeling in which Ang-2 blocks a constitutive stabilizing action of Ang-1, resulting in less stable pericyte/endothelial cell interactions and consequent reversion of blood vessels to a more plastic state for initiation or regression of angiogenesis[9,164]. Indeed, studies of rat tumor models have provided evidence that Ang2 induces angiogenic sprouting in the presence of VEGF and stimulates vessel regression in the absence of VEGF[171,172].

Surprisingly, a recent study of Ang2 null mice has revealed that the angiopoietins are likely to play a role in lymphatic development as well[173]. While defective postnatal angiogenesis is observed in Ang-2-deficient mice, these animals also exhibit gross abnormalities of the lymphatic system, displaying lymphedema, chylous ascities, and structurally irregular and leaky lymphatic vessels. Interestingly, Ang1 overexpression in Ang2-deficient mice is sufficient to rescue the lymphatic but not blood vascular phenotype, adding additional complexity to the relative functions of Ang1 and Ang2 and their potential interplay with other lymphangiogenic factors[2,173].

## 7.1 Other Molecules Implicated in Lymphatic Development and Function

### 7.1.1 LYVE-1

LYVE-1 was originally identified by its strong homology to CD44; it serves as a lymphatic endothelium-specific hyaluronan receptor[174]. Hyaluron, a mucopolysaccharide polymer that comprises a significant component of the extracellular matrix, undergoes constant turnover and is partially degraded within the lymphatic system[2,175,176]. Hyaluronan turnover increases during tissue injury and its breakdown products induce inflammatory responses, resulting in angiogenesis, chemokine production, and dendritic cell recruitment. While LYVE-1 is expressed by various cell types such as activated macrophages and sinusoidal endothelium of the liver and spleen, it is generally absent in endothelia from blood vasculature and serves as one of the most important lymphatic-specific markers, along with Prox1 and VEGFR-3. Surprisingly, LYVE-deficient mice appear normal, and no obvious lymphatic vascular malfunctions or morphological abnormalities have been detected thus far[2].

### 7.1.2 Podoplanin

Podoplanin, a mucin-type transmembrane protein, is expressed by lymphatic but not by blood vascular endothelial cells[2,58,60,63,177,178]. Like LYVE-1, podoplanin is expressed by various other cell types including type I alveolar cells of the lung, cells of the choroid plexus, and kidney podocytes[178,179]. During murine embryonic development, podoplanin is absent in the vascular system until E11.5-E12.5, when it becomes expressed in all cardinal vein endothelial cells, including the budding Prox-1 positive, lymphatically committed cells[178-180]. As with LYVE-1 and VEGFR-3, podoplanin expression continues in these lymphatically-committed cells but is progressively down-regulated in venous endothelial cells[178]. By the time of birth, podoplanin expression is generally restricted to lymphatic endothelial cells[59].

Podoplanin-deficient mice display cutaneous lymphedema associated with impaired lymphatic transport[178], dilated lymphatic vessels of the intestine and skin, and neonatal lethality due to lung failure[178,179]. Because podoplanin null mice die at birth, the generation of tissue-specfic knockout mice will be important for the elucidation of the role of podoplanin in postnatal lymphatic development[181]. *In vitro*, podoplanin appears to play an important role in lymphatic endothelial cell migration, adhesion and tube formation[178].

### 7.1.3 Neuropilins

Neuropilins are widely expressed, multifunctional nonkinase type I transmembrane proteins that serve as receptors for class III semaphorins, which are known to mediate axon guidance[182]. Neuropilins also modulate angiogenesis[182], binding to several VEGF family members[142]. Because neuropilins are believed to be non-signaling, they are thought to require the presence of a signal-transducing receptor in order to mediate their effects[9]; by binding to various VEGFs, neuropilins may act as co-receptors for the VEGFRs, enhancing the efficiency of their tyrosine kinase signals[183].

Neuropilin-2 serves as a receptor for VEGF-A165 (a VEGF isoform), PlGF-2 and VEGF-C, and is expressed primarily in veins and visceral lymphatic vessels[2,182]. Endothelial neuropilin-1 expression in chick embryos is restricted primarily to arteries, suggesting that the neuropilins may play a role in determining endothelial arterial *versus* venous identity[9,184]. Neuropilin-2-deficient mice manifest neural defects but also demonstrate a relative paucity of small lymphatic vessels during development, whereas arteries and veins are not affected[85]. This suggests a selective requirement of neuropilin-2 in lymphatic development[9]. The observation that neuropilin-2 can bind VEGF-C suggests that neuropilin-2 may be involved in VEGFR-3 signaling in lymphatics[113,185].

## 8. LYMPHANGIOGENESIS IN HEALTH AND DISEASE

Lymphangiogenesis is likely to represent an integral feature of tissue repair and inflammatory reactions in most organs[9]. During wound healing, for instance, VEGFR-3 positive lymphatic cells have been shown to bud out from pre-existing lymphatics into the granulation tissue[186,187]. The need for interstitial fluid drainage may explain the apparent requirement for lymphatic growth in the setting of angiogenesis, in which leaky, immature blood vessels are present[9]. Primary impairment of lymphatic function results in lymphedema, a disease state characterized by disfiguring swelling of the extremities[188]. Conversely, abnormal proliferation of lymphatic endothelial cells is seen in lymphangiomas and lymphangiosarcomas, and possibly in Kaposi's sarcoma[189]. Lymphatic vessels further serve as likely routes for tumor metastasis.

Until recently, there has been little delineation of the molecular mechanisms that are invoked in lymphatic disease. Several gene mutations have now been identified as causal for human lymphedema, the first lymphatic markers and lymphangiogenic growth factors have been characterized, and animal models have facilitated the development of therapeutic applications for both heritable and acquired forms of lymphedema. Animal models have further been used to analyze mechanisms of tumor metastasis and to test strategies for the inhibition of metastasis[190]. These novel developments now afford us the ability to modulate the lymphangiogenic process and develop targeted therapies for diseases associated with lymphatic dysfunction.

## 9. LYMPHEDEMA

Impairment of lymphatic function characterizes a pathologic condition termed lymphedema, in which inadequate transport of interstitial fluid, edema, impaired immunity and fibrosis are observed[191]. These changes are frequently accompanied by disfiguring and disabling swelling of the extremities[191]. The full pathologic expression of lymphedema is complex and, to a large degree, poorly understood. According to the classical model, an imbalance between lymphatic load and transport capacity predicates the accumulation of protein and obligate fluid in the interstitial space[191-193]. This leads to further edema formation, in part as a consequence of increases in tissue colloid osmotic pressure. Ensuing architectural changes are often profound[191,194-196]. Chronic lymph stasis predisposes to an increase in the number of fibroblasts, adipocytes, and keratinocytes in the edematous tissues. Mononuclear cells often demarcate the inflammatory response[194,195]. Ultimately, skin thickening and subcutaneous fibrosis ensue[191], although the mechanisms of this transformation are still not well understood.

Lymphedema can occur as a manifestation of heritable malformations of the lymphatics or as a result of infectious, traumatic, or postsurgical disruption of these circulatory structures[188,191-193,196]. A useful classification of lymphedema differentiates between primary and secondary etiologies. Primary lymphedemas, rare developmental disorders which are often classified by the age at which edema first appears, have three recognized forms: congenital, which is apparent within the first two years of life; praecox, which usually appears at the onset of puberty; and tarda, which typically appears after 35 years of age.

Recent investigation into the most common, autosomal dominant form of congenital primary lymphedema, Milroy's disease, has provided strong evidence that this condition results from missense mutations in the VEGFR3 receptor; these putatively inactivate the tyrosine kinase signaling mechanism required for lymphatic development[129,193]. Secondary lymphedema, which is much more prevalent than the primary form, develops after disruption of lymphatic pathways by regional trauma or infection, or as a consequence of surgery or radiotherapy[191-193]. Edema of the arm after breast cancer interventions is probably the most common cause of lymphedema in the United States; the global incidence of secondary incidence of secondary lymphedema can be ascribed predominantly to filiarisis, which afflicts more than 90 million people[192].

Without regard to the specific pathogenesis, lymphedema is a chronic, unrelenting condition that predisposes to substantial morbidity and loss of function[191-193]. In the setting of breast cancer alone, postsurgical lymphatic dysfunction is estimated to affect approximately 30% of the estimated 2 million American breast cancer survivors[192]. Lymphedema lacks a cure, posing long-term physical and psychological difficulties for the patient and a complex therapeutic challenge for the physician[191,192,197-201]. Current treatment modalities are limited to physiotherapeutic interventions that reduce edema volume but, at best, provide only partial relief to afflicted individuals[191,192,202]. Furthermore, such conservative measures do not guarantee longterm freedom from the advent of ostensibly irreversible fibrosis[192].

For these reasons, there has been substantial interest in the emerging field of growth factor- and gene therapy-mediated lymphangiogenesis. In analogy to proposed angiogenic treatment strategies for diseases of the coronary and peripheral blood vasculature[203-205], molecular approaches may ultimately provide a therapeutic window for reversing the stigmata of both primary and secondary lymphatic insufficiency[193,196]. Initial experimental observations indicate that gene- and growth factor-mediated therapeutic lymphangiogenesis with VEGFC holds promise for the treatment of both primary and secondary forms of lymphedema[185,196].

## 10. GENETIC ALTERATIONS IN LYMPHATIC DISEASE

Primary hereditary lymphedema can occur as an autosomal dominant condition where lymphedema is the only clinically apparent abnormality (Milroy's disease), or as a condition where lymphedema occurs as one manifestation of a more complex genetic syndrome, with either an autosomal recessive or autosomal dominant pattern of inheritance[154]. The clinical heterogeneity of lymphedema syndromes is being confirmed by the emerging molecular genetic studies of these phenotypes.

Milroy's disease has been reported to be linked in several families to the FLT4 locus in the distal chromosome 5q, encoding vascular VEGFR3[206-208]. In affected family cohorts, missense mutations of the VEGFR3 gene yield an inactive tyrosine kinase with presumed defects in downstream signaling[129]. The inherited disease is now believed to impair the ability of the VEGFR3 receptor to orchestrate lymphangiogenesis in development. When these mutant receptors are overexpressed in heterologous cells, they fail to demonstrate ligand-dependent autophosphorylation in response to VEGF-C, and are less effective than wild-type receptors in eliciting downstream signaling *in vitro*[129]. It is thought that enhanced stability of the mutant form of the receptor on the cell surface is responsible for this blunting of signaling[209,210]. Interestingly, humans with chromosomal abnormalities leading to heterozygous deletion of the VEGFR3 gene do not display lymphedema as a phenotype[211,212], suggesting that lymphedema is not due to haploinsufficiency for VEGFR3[154].

While mutations inhibiting the biologic activity of VEGFR-3 are causes of primary lymphedema, only about 5% of patients with primary lymphedema carry VEGFR3 mutations[9]. Other genetic loci have been implicated in certain families with Milroy's disease and other lymphedema syndromes. For instance, the more prevalent mutations in the FOXC2 gene result in the hereditary lymphedema-distichiasis syndrome[213]. Lymphedema-distichiasis is an autosomal dominant syndrome characterized by the onset of lymphedema at puberty. Most affected individuals have distichiasis, an extra row of eyelashes arising inappropriately from the meibomian glands[214]. Expression of other features of the lymphedema distichiasis syndrome (congenital heart disease, ptosis, cleft lip/palate, venous malformations) is variable, both in severity and frequency of occurrence[154,215-220].

Lymphedema-distichiasis has been linked in numerous families to the FOXC2 locus in chromosome 16q[213,221-225]. FOXC2 is a member of the forkhead family of transcription factors[226], and participates in a wide variety of developmental processes during embryogenesis[154]. Of the more than 30 FOXC2 mutations in lymphedema-distichiasis that have been characterized, all except one is predicted to lead to a truncation of the wildtype protein[223-225]. Neither the mutation site nor the protein truncation point clusters among families. Notably, the lack of correlation between the truncation point and the occurrence of specific phenotypic features suggests that FOXC2 may be a dosage-sensitive gene and that there is a haploinsufficiency effect of FOXC2 for lymphedema-disthichiasis[154].

FOXC2 null mice die pre- or perinatally, displaying lethal heart defects and skeletal abnormalities[227,228], while mice heterozygous for a null allele at the FOXC2 locus appear to have multiple anterior segment abnormalities[229]. The lymphatic phenotypes in both the homozygous and heterozygous null animals will require further investigation. Other studies involving genetically modified FOXC2 mice have not yet identified a clear role of this gene in lymphatic development[154].

## 11. ANIMAL MODELS FOR THE STUDY OF LYMPHATIC DISEASE

Experimental models of both primary and secondary lymphedema now provide us with tools to develop and test new therapies for lymphatic dysfunction. The prototypic, successful approach to experimental, acquired lymphedema was developed in canine hindlimb[193,230]. Chronic, resistant lymphedema can be obtained if, in addition to transection of the lymphatics, a circular strip of skin, subcutaneous tissues, fascia, and periosteum is removed from the thigh so as to interpose a scar to close the capillary network that would otherwise retain the ability to restore lymph flow. Analogous surgical approaches have been adapted to smaller laboratory animals, capitalizing on lower costs and reduction of the time lapse required for clinically relevant observations[231]. Models of postsurgical lymphedema in the rat hindlimb and mouse tail are examples[232-236]. An inherent negative characteristic of surgical attempts to generate experimental chronic lymphedema is the substantial propensity to surgical morbidity that is out of proportion to the human clinical counterpart[193]. The regenerative capacity of the lymphatics[237], development of collateral circulations[238], and formation of lymphaticovenous shunts serve as further obstacles in creating a sustained, stable post-surgical model of lymphatic insufficiency.

Many of these problems have been circumvented through the use of the rabbit ear, a structure that has the advantage of providing a relatively large area of homogenous tissue that receives lymphatic drainage through a small conduit which is easily accessed for surgical intervention[193]. In a recently described rabbit ear model[196] that utilizes a modification of a previously described approach[239], a circumferential strip of skin, subcutaneous tissue, and perichondrium is excised from the base of the ear, with protection of the central neurovascular bundle and preservation of the chondrium. Major lymphatic channels are then resected and ligated. Stable, chronic lymphedema that appears to simulate the human condition ensues, as demonstrated by lymphoscintigraphic findings of impaired lymphatic transport and light microscopic evidence of cutaneous changes that are observed in human chronic lymphedema.

While surgical models of lymphatic insufficiency now provide a means of developing potential therapies for acquired lymphedema, genetic animal models have permitted the study of inheritable lymphatic insufficiency. Two existing mouse strains[119,185] demonstrate phenotypes of primary lymph-edema. A transgenic mouse has been described[119] that, at least transiently, has attributes of the human disease[193,231]. The model of Makinen et al. expresses a chimeric protein consisting of the ligand-binding domain of VEGFR3 and an IgG Fc domain. Overexpression of this soluble receptor in the skin competes for VEGF-C/VEGF-D binding with the endogenous receptor, leading to inhibition of fetal lymphangiogenesis and regression of developing lymphatic vessels. Importantly, there is no effect on the blood vasculature. The mice develop a lymphedema-like phenotype with edema of the limbs, increased deposition of subcutaneous fat, and dermal fibrosis. Initially, there is a loss of lymphatic tissue in internal organs and lack of dermal lymphatic vessel development although, over time, there is an escape from this pattern of biological expression.

Of potentially greater relevance[231] to Milroy's disease is the Chy mouse model[185]. The Chy mouse, whose phenotype maps to chromosome 11, is characterized by a phenotype of chylous ascites, hypoplastic cutaneous lymphatic vessels[185] and, like the transgenic mouse, a form of limb edema that bears a superficial resemblance to human primary lymphedema[193,231]. However, the pathology is even more tightly congruent because, in this mouse line, the VEGFR3 receptor is mutant. Specifically, the Chy mouse contains a heterozygous A3157T mutation that results in an I1053F substitution within a highly conserved catalytic domain of the VEGFR3 receptor, in close proximity to the locus of VEGFR3 mutations in human primary lymphedema[185]. This mutation inactivates the receptor tyrosine kinase, a finding consistent with the tyrosine kinase-inactive VEGFR-3s seen in Milroy's disease[129]. Congenital lymphedema has been described in dog, pig, and cattle, but none of these animal models have been studied at the molecular[154].

## 12. INSIGHTS INTO THE TARGETED TREATMENT OF LYMPHEDEMA

The identification of specific genes and signaling cascades involved in the regulation of lymphatic vessel development has established a promising basis for the development of targeted treatments for lymphatic dysfunction. Indeed, in analogy to proposed proangiogenic treatment strategies for circulatory insufficiency of the myocardium and other major vascular beds[203-205,240-243], molecular approaches may soon provide a therapeutic window for reversing the stigmata of both primary and secondary lymphatic insufficiency[193,231].

To date, VEGF-C appears to be the most promising molecular tool for the treatment of primary and secondary lymphedema in animal models. In the Chy mouse model that closely resembles Milroy's disease, the local delivery of VEGF-C by adeno-associated virus-mediated delivery of recombinant human VEGF-C promotes lymphangiogenesis and an amelioration of lymphedema that accompanies the generation of new, functional lymphatic vasculature[185,231]. Furthermore, crosses between Chy and K14-VEGF-C156S mice, which overexpress the VEGFR3-specific ligand VEGF-C156S[118], leads to restored lymphatic function in the Chy X K14-VEGF-C156S offspring[185]. These observations indicate that delivery of an excess of ligand can restore normal lymphatic function in the setting of impaired VEGFR3 signaling[154].

VEGF-C has also been shown to have potential applicability for the treatment of secondary lymphatic insufficiency. In a study of growth factor-mediated therapeutic lymphangiogenesis in acquired post-surgical lymphedema[196], a single dose of human recombinant VEGF-C was administered to the rabbit ear in which lymphatics had been previously surgically ablated. This therapeutic intervention was sufficient to induce lymphangiogenesis, improve dynamic lymphatic function, and reverse the tissue hypercellularity that characterizes the untreated lymphedematous state. The demonstration of successful lymphangiogenesis in this model enhances the prospects of growth factor-mediated therapies for a larger segment of the disease population than is represented by Milroy's disease alone. However, it is important to recognize that, as tumor-induced lymphangiogenesis has been associated with enhanced lymph node metastasis[244-249], the risk of enhanced growth and spread of tumors during VEGF-C therapy needs careful evaluation[9,196].

The vehicle by which VEGF-C is administered in therapeutic lymphangiogenesis carries important implications for the durability of the therapeutic response. For instance, the half-life of VEGF-C in the blood circulation is short[118]; furthermore, the long-term efficacy of direct growth factor therapy remains to be established. In contrast, while recombinant adenoviruses are efficient gene transfer vectors, transgene expression is quickly lost due to a mounted immune response to viral proteins of the vector[250]; intradermal adenovirus-mediated VEGF-C gene transfer in the mouse skin has been shown to result in a strong lymphangiogenic response[146,250], but the majority of the newly formed lymphatic vessels regress when the adenovirus is no longer active.

Whereas adenoviral gene transfer appears to yield only short-term expression, adeno-associated viruses (AAVs) yield transgene expression that may last for over a year[250,251]. AAVs are non-pathogenic human viruses that do not elicit inflammatory reactions or cytotoxic immune responses[252]. It appears that the low transgene expression levels and slow kinetics of AAV-VEGF-C infection are factors that predispose to a more controlled lymphangiogenesis than attained from adenoviral-mediated VEGF-C expression[250]. Indeed, AAV-mediated expression of VEGF-C in Chy mice results in long-lasting transgene expression and sustained, VEGF-C-induced cutaneous lymphatic vessels[253]. As recombinant AAVs have also been shown to result in long-term gene expression in humans, AAVs may ultimately prove to be more suitable for lymphangiogenic therapy in humans[250].

Proangiogenic gene therapy with VEGF, while capable of inducing angiogenesis, carries the potential for significant side effects. For instance, the vessels that VEGF helps to create are immature, tortuous and leaky, and often lack perivascular support structures[9,254,255]. Edema induced by VEGF overexpression further complicates VEGF-mediated neovascularization, although recent evidence suggests that this can be avoided by providing Ang-1 for vessel stabilization[168,256]. Analogously, lymphangiogenic therapy with VEGF-C has also been shown to be capable of promoting vascular permeability[141], a potentially undesirable side-effect in the treatment of human lymphedema.

A recent study has demonstrated that adenoviral- or AAV-mediated VEGF-C overexpression in mouse skin, in addition to inducing lymphangiogenesis, also promotes blood vessel enlargement, tortuosity and leakiness, though no frank angiogenesis is noted[141]. Mechanistically, VEGF-C has been shown to stimulate the release of endothelial nitric oxide, a potent vasodilator, which may partially account for the described enhancement in vascular permeability[257]. The blood vascular permeability effects observed with adenoviral- and AAV-VEGF-C may furthermore result from VEGFR-2 mediated vasodilation[9], as the veins and venules of the skin were shown to express VEGFR2[141].

In the same study, it was shown that adenoviral Ang-1 is able to reverse the blood vessel leakiness induced by VEGF-C, while not affecting lymphangiogenesis[141]. Ang-1 has previously been reported to protect adult blood vasculature against the plasma leakage induced by VEGF or inflammatory mediators[168,256]. And, as described earlier, mouse models have suggested that Ang-1 signaling via Tie-2 is required for the stabilization of blood vessels, and for the maintenance of interactions between endothelial cells and the surrounding extracellular matrix and mesenchyme[160,167,258]. It is conceivable that angiopoietins may play similar stabilizing roles in the development of large lymphatic collecting vessels[9], which usually have tightly associated smooth muscle cells but display loosely organized smooth muscle cell coverage in the absence of Ang-2[173].

To develop an approach to therapeutic lymphangiogenesis which circumvents possible VEGF-C-induced permeability effects on the blood vasculature, investigations have explored the potential of VEGF-C156S, a VEGFR-3-specific mutant (so as to avoid potential VEGF-C interactions with VEGFR2), as a therapeutic agent in lymphedema[253]. VEGF-C156S, when expressed as a transgene in mouse skin, promotes lymphatic vessel hyperplasia, whereas VEGF-C induces an increase in the number of lymphatic vessels[253], suggesting that VEGFR-2 activation is required for the induction of lymphatic sprouting in embryogenesis[9]. Viral administration of VEGF-C156S in adult skin, however, is capable of inducing the formation of new lymphatic sprouts, though this is less pronounced than with native VEGF-C[253]. Importantly, administration of adenoviral VEGF-C156S does lead to the blood vascular permeability effects seen with VEGF-C adenoviruses, carrying important implications for the development of future gene therapies for human lymphedema.

## 13. TUMOR LYMPHANGIOGENESIS AND VEGF-C/D

Tumorigenesis and tumor metastasis are multi-step processes[9]. Many factors, including proteolytic and migratory activity of tumor cells, expression of adhesion molecules, and extracellular matrix deposition are known to contribute to tumor growth and metastasis. The angiogenic switch is additionally recognized as one of the key events in tumorigenesis[259,260]. In fact, the ability of tumor cells to induce angiogenesis is considered a prerequisite for tumor growth, invasion, and successful metastasis. The importance of the lymphatic system as a pathway for metastasis is likewise well recognized. For instance, the extent of lymph node involvement in cancers is a key prognostic factor in patient outcome. However, while there is a substantial body of experimental work addressing the role of hematogenous spread in tumor dissemination[259,261], the mechanisms by which tumor cells interact with the lymphatics still require elucidation[260].

In principle, tumor cells can invade pre-existing lymphatic vessels directly, or enter new lymphatic vessels that have been formed by tumor-induced lymphangiogenesis[9]. It is currently unclear whether lymphangiogenesis is an integral part of tumorigenesis in humans and whether activation of the lymphatic system is restricted to specific cancers and/or tumor stages[260]. Historically, few studies had addressed the issue of whether tumor lymphangiogenesis occurs at all, and only recently have novel molecular markers for the lymphatics provided evidence for tumor lymphangiogenesis. Functional lymphatics containing clusters of tumor cells are often detected in the tumor periphery[262]. While no evidence for intratumoral lymphatics has been found in human melanomas, cervical, ovarian and liver carcinomas, and in experimental models of pancreatic cancer and melanoma[244,263-268], intratumoral lymphatic channels have been observed in autochthonous human breast cancers, head and neck squamous cell carcinomas and in numerous experimental tumor models[49,245-247,249,263,269-272].

The biological significance of intratumoral lymphatic vessels and their role in tumor dissemination remains debated[260,262,273]. Lymphatic vessels identified in tumors with lymphatic-specific molecular markers have not been confirmed by lymphangiography, a technique in which labeled macromolecules are injected into the interstitium for uptake into the lymphatics. It has therefore been suggested that intratumoral lymphatics are nonfunctional and that tumor cells cannot utilize these lymphatics for transport to lymph nodes. It has alternatively been proposed that the absence of detectable perfusion in intratumoral lymphatics is not due to their functional impairment but rather to their collapse under interstitial pressure[273] induced by growing cancer cells. It has further been proposed that, even if intratumoral lymphatics are not fully functional with respect to fluid uptake, they might still promote metastatic tumor spread by creating increased opportunities for tumor cells to leave the primary tumor site. It is conceivable, for instance, that growth factor-mediated activation of intratumoral lymphatics might induce lymphatic endothelial cells to facilitate active tumor cell entry into the lymphatics[260].

Whereas the lymphatic system was once thought to play a passive role in the metastatic process, there is now an increasing body of evidence suggesting that lymphatic vasculature plays an active role in promoting tumor metastasis, and that VEGF-C may be central to this role[245,260]. There is abundant evidence for the expression of VEGF-C in human tumors. VEGF-C is expressed in breast, colon, gastric, lung, squamous cell and thyroid cancers, in addition to mesotheliomas and neuroblastomas[274-285]. Moreover, a correlation between VEGF-C expression and rate of metastasis to lymph nodes has been found in breast, colorectal, gastric, lung, prostate and thyroid cancers[274-280,282,284,286].

Experimental animal models have supported the role of VEGF-C in promoting tumor lymphangiogenesis and lymph node metastasis. In one study[244], transgenic (RipVEGF-C) mice that express VEGF-C specifically in pancreatic beta-cells and that, accordingly, develop a lymphatic network around the beta cells, were mated with a second transgenic (Rip1Tag2) strain that characteristically develop non-metastatic pancreatic beta cell tumors. The double transgenic mice displayed VEGF-C induced lymphangiogenesis around the beta cell tumors, in addition to metastatic spread of tumor cells to pancreatic and regional lymph nodes. The increased amount of peritumoral lymphatics seen in the double transgenic strain supports the notion that VEGF-C-induced increases in peritumoral lymphatic vessels makes these vessels more accessible to tumor cells[9].

The functional importance of VEGF-C and lymphangiogenesis for tumor progression has also been assessed in orthotopic tumor models in which human breast cancer cells overexpressing VEGF-C are implanted into immunosuppressed mice[245,247]. In one model[245], VEGF-C expression has been shown to increase both peritumoral and intratumoral vessel density, and to enhance tumor metastasis to regional lymph nodes and lung. The degree of tumor metastases appears to correlate with intratumoral lymphatic vessel density as well as with the depth of lymphatic vessel invasion into the tumors. VEGF-D has similarly been shown to promote the metastatic spread of tumor cells in a mouse 293EBNA tumor model[246].

Studies have demonstrated that VEGF-C-induced tumor lymph-angiogenesis and lymph node metastasis in mice can be inhibited by adenoviral expression of a soluble VEGFR-3, which competes for VEGF-C/VEGF-D binding with their endogenous receptors[247,248]. In another mouse tumor model[287], inactivation of VEGFR-3 with a blocking monoclonal antibody was sufficient to suppress tumor growth. However, VEGFR-3 inactivation in this tumor model surprisingly did not affect lymphatic vasculature, but inhibited the neo-angiogenesis of tumor-bearing tissues.

The role of angiogenesis during VEGF-C and VEGF-D mediated tumorigenesis remains unclear. Angiogenesis is not induced by VEGF-C overexpression in the orthotopic breast cancer models described above[245,247]. However, VEGF-C has been shown to induce both tumor lymph-angiogenesis and angiogenesis in a human malignant xenoplant model[263], and VEGF-D has been shown to promote tumor angiogenesis and growth in the mouse 293EBNA tumor model[246]. The differences in tumor angiogenic properties of VEGF-C/D in various studies may reflect differences in proteolytic processing of these growth factors in various tumor types[9].

In humans, tumor lymphangiogenesis has been found to be highly associated with enhanced metastatic potential of human malignant melanomas, and may serve as a novel prognostic factor for tumor metastasis[288]. Nevertheless, the definitive role of VEGF-C and VEGF-D expression in promoting lymphangiogenesis in human tumors, and whether this leads to higher propensity for metastasis, remains to be established. Although preliminary results from animal models and patient studies have suggested that VEGF-C and VEGF-D have important implications in tumor metastasis, further studies are required to establish their importance in human cancers. As described earlier, numerous reports have suggested a correlation between VEGF-C expression and lymphatic metastasis in human tumors, while less is known about VEGF-D in human tumors[109,274]. While it is apparent that the effects of VEGF-C/D and receptor signaling on tumor progression require further investigation, the above observations offer promise for the development of targeted molecular therapies for inhibiting lymphatic metastasis.

## 14. CONCLUSIONS

Although the lymphatic vascular system plays a vital role in mediating tissue fluid homeostasis, immune responses, and tumor dissemination, only recently have the mechanisms of lymphatic development and its association with disease been elucidated at the molecular level. The discovery and characterization of lymphatic-specific growth factors, receptors, and transcriptional regulators have revealed that lymphangiogenesis shares parallels with vasculo/angiogenesis, both in embryonic development and in post-natal life. The identification of signaling cascades involved in the regulation of lymphatic vessel development have established a promising basis for the development of targeted treatments for lymphatic dysfunction. However, initial animal studies have demonstrated that growth factor-mediated therapeutic lymphangiogenesis carries the potential for significant side effects, as also noted in studies of angiogenic treatment strategies for coronary disease. The possible overlapping functions of various lymphangiogenic growth factors and receptors (vis-à-vis lymphangiogenesis versus blood vascular angiogenesis), and their potential for interaction adds further complexity to their roles *in vivo*. Further investigation will prove useful in developing specific, targeted molecular therapies for the treatment of human lymphatic disease.

## REFERENCES


1. Szuba A, Shin WS, Strauss HW, Rockson S. The third circulation: radionuclide lymphoscintigraphy in the evaluation of lymphedema. J Nucl Med. 2003; 44:43-57.
2. Hong YK, Shin JW, Detmar M. Development of the lymphatic vascular system: a mystery unravels. Dev Dyn. 2004; 231:462-73.
3. Leak LV, Burke JF. Fine structure of the lymphatic capillary and the adjoining connective tissue area. Am J Anat. 1966; 118:785-809.
4. Leak LV. Electron microscopic observations on lymphatic capillaries and the structural components of the connective tissue-lymph interface. Microvasc Res. 1970; 2:361-91.
5. Casley-Smith JR. The fine structure and functioning of tissue channels and lymphatics. Lymphology. 1980; 13:177-83.

6. Barsky SH, Baker A, Siegal GP, Togo S, Liotta LA. Use of anti-basement membrane antibodies to distinguish blood vessel capillaries from lymphatic capillaries. Am J Surg Pathol. 1983; 7:667-77.
7. Ezaki T, Matsuno K, Fujii H, Hayashi N, Miyakawa K, Ohmori J, Kotani M. A new approach for identification of rat lymphatic capillaries using a monoclonal antibody. Arch Histol Cytol. 1990; 53 Suppl:77-86.
8. Oh SJ, Jeltsch MM, Birkenhager R, McCarthy JE, Weich HA, Christ B, Alitalo K, Wilting J. VEGF and VEGF-C: specific induction of angiogenesis and lymphangiogenesis in the differentiated avian chorioallantoic membrane. Dev Biol. 1997; 188:96-109.
9. Lohela M, Saaristo A, Veikkola T, Alitalo K. Lymphangiogenic growth factors, receptors and therapies. Thromb Haemost. 2003; 90:167-84.
10. Olszewski WL. Lymphatics, lymph and lymphoid cells: an integrated immune system. Eur Surg Res. 1986; 18:264-70.
11. Cornford ME, Oldendorf WH. Terminal endothelial cells of lymph capillaries as active transport structures involved in the formation of lymph in rat skin. Lymphology. 1993; 26:67-78.
12. Lubach D, Ludemann W, Berens von Rautenfeld D. Recent findings on the angioarchitecture of the lymph vessel system of human skin. Br J Dermatol. 1996; 135:733-7.
13. Kubik S, Manesta M. Anatomy of the lymph capillaries and precollectors of the skin. In: Bollinger A, Partsch H, Wolfe J, eds. The Initial Lymphatics. Stuttgart, Germany: Thieme-Verlag; 1985:66-74.
14. Moghimi SM, Bonnemain B. Subcutaneous and intravenous delivery of diagnostic agents to the lymphatic system: applications in lymphoscintigraphy and indirect lymphography. Adv Drug Deliv Rev. 1999; 37:295-312.
15. Nerlich AG, Schleicher E. Identification of lymph and blood capillaries by immunohistochemical staining for various basement membrane components. Histochemistry. 1991; 96:449-53.
16. Pfister G, Saesseli B, Hoffmann U, Geiger M, Bollinger A. Diameters of lymphatic capillaries in patients with different forms of primary lymphedema. Lymphology. 1990; 23:140-4.
17. Ikomi F, Hanna G, Schmid-Schonbein GW. Intracellular and extracellular transport of perfluoro carbon emulsion from subcutaneous tissue to regional lymphatics. Artif Cells Blood Substit Immobil Biotechnol. 1994; 22:1441-7.
18. Ikomi F, Hanna GK, Schmid-Schonbein GW. Mechanism of colloidal particle uptake into the lymphatic system: basic study with percutaneous lymphography. Radiology. 1995; 196:107-13.
19. Ikomi F, Schmid-Schonbein GW. Lymph transport in the skin. Clin Dermatol. 1995; 13:419-27.
20. O'Morchoe PJ, Yang VV, O'Morchoe CC. Lymphatic transport pathways during volume expansion. Microvasc Res. 1980; 20:275-94.
21. Porter CJ, Charman SA. Lymphatic transport of proteins after subcutaneous administration. J Pharm Sci. 2000; 89:297-310.
22. Leak LV. Lymphatic removal of fluids and particles in the mammalian lung. Environ Health Perspect. 1980; 35:55-76.
23. Castenholz A. Structure of initial and collecting lymphatic vessels. In: Olszewski W, ed. Lymph Stasis: Pathophysiology, Diagnosis and Treatment. Boca Raton, FL: CRC Press; 1991:15-41.

24. Wenzel-Hora BI, Berens von Rautenfeld D, Majewski A, Lubach D, Partsch H. Scanning electron microscopy of the initial lymphatics of the skin after use of the indirect application technique with glutaraldehyde and MERCOX as compared to clinical findings. Lymphology. 1987; 20:126-44.
25. Olszewski W, Machowski Z, Sokolowski J, Wojciechowski J. Alterations in lymphatic vessels in the course of chronic experimental lymphedema. Pol Med J. 1970; 9:1441-8.
26. Schmid-Schonbein GW. Microlymphatics and lymph flow. Physiol Rev. 1990; 70:987-1028.
27. Eisenhoffer J, Kagal A, Klein T, Johnston MG. Importance of valves and lymphangion contractions in determining pressure gradients in isolated lymphatics exposed to elevations in outflow pressure. Microvasc Res. 1995; 49:97-110.
28. Wilting J, Neeff H, Christ B. Embryonic lymphangiogenesis. Cell Tissue Res. 1999; 297:1-11.
29. Cursiefen C, Chen L, Dana MR, Streilein JW. Corneal lymphangiogenesis: evidence, mechanisms, and implications for corneal transplant immunology. Cornea. 2003; 22:273-81.
30. Olszewski W. Lymphology and the lymphatic system. In: Olszewski W, ed. Lymph Stasis: Pathophysiology, Diagnosis and Treatment. Boca Raton, FL: CRC Press; 1991:4-12.
31. Tedder TF, Steeber DA, Chen A, Engel P. The selectins: vascular adhesion molecules. Faseb J. 1995; 9:866-73.
32. Butcher EC, Williams M, Youngman K, Rott L, Briskin M. Lymphocyte trafficking and regional immunity. Adv Immunol. 1999; 72:209-53.
33. Rosen SD. Endothelial ligands for L-selectin: from lymphocyte recirculation to allograft rejection. Am J Pathol. 1999; 155:1013-20.
34. Kunkel EJ, Butcher EC. Chemokines and the tissue-specific migration of lymphocytes. Immunity. 2002; 16:1-4.
35. Gunn MD, Tangemann K, Tam C, Cyster JG, Rosen SD, Williams LT. A chemokine expressed in lymphoid high endothelial venules promotes the adhesion and chemotaxis of naive T lymphocytes. Proc Natl Acad Sci U S A. 1998; 95:258-63.
36. Cyster JG. Chemokines and the homing of dendritic cells to the T cell areas of lymphoid organs. J Exp Med. 1999; 189:447-50.
37. Muller A, Homey B, Soto H, Ge N, Catron D, Buchanan ME, McClanahan T, Murphy E, Yuan W, Wagner SN, Barrera JL, Mohar A, Verastegui E, Zlotnik A. Involvement of chemokine receptors in breast cancer metastasis. Nature. 2001; 410:50-6.
38. Wiley HE, Gonzalez EB, Maki W, Wu MT, Hwang ST. Expression of CC chemokine receptor-7 and regional lymph node metastasis of B16 murine melanoma. J Natl Cancer Inst. 2001; 93:1638-43.
39. Risau W, Flamme I. Vasculogenesis. Annu Rev Cell Dev Biol. 1995; 11:73-91.
40. Risau W. Mechanisms of angiogenesis. Nature. 1997; 386:671-4.
41. Yancopoulos GD, Klagsbrun M, Folkman J. Vasculogenesis, angiogenesis, and growth factors: ephrins enter the fray at the border. Cell. 1998; 93:661-4.
42. Asellius G. De lactibus sive lacteis venis. In. Milan: Mediolani; 1627.
43. Sabin F. On the origin of the lymphatic system from the veins and the development of the lymph hearts and thoracic duct in the pig. Am J Anat. 1902; 1:367-391.
44. Sabin F. On the development of the superficial lymphatics in the skin of the pig. Am J Anat. 1904; 3:183-195.
45. Huntington G, McClure CT. he anatomy and development of the jugular lymph sac in the domestic cat (Felis domestica). Anat Rec. 1908; 2:1-19.

46. Kaipainen A, Korhonen J, Mustonen T, van Hinsbergh VW, Fang GH, Dumont D, Breitman M, Alitalo K. Expression of the fms-like tyrosine kinase 4 gene becomes restricted to lymphatic endothelium during development. Proc Natl Acad Sci U S A. 1995; 92:3566-70.
47. Dumont DJ, Jussila L, Taipale J, Lymboussaki A, Mustonen T, Pajusola K, Breitman M, Alitalo K. Cardiovascular failure in mouse embryos deficient in VEGF receptor-3. Science. 1998; 282:946-9.
48. Wigle JT, Oliver G. Prox1 function is required for the development of the murine lymphatic system. Cell. 1999; 98:769-78.
49. Wigle JT, Harvey N, Detmar M, Lagutina I, Grosveld G, Gunn MD, Jackson DG, Oliver G. An essential role for Prox1 in the induction of the lymphatic endothelial cell phenotype. Embo J. 2002; 21:1505-13.
50. Schneider M, Othman-Hassan K, Christ B, Wilting J. Lymphangioblasts in the avian wing bud. Dev Dyn. 1999; 216:311-9.
51. Papoutsi M, Tomarev SI, Eichmann A, Prols F, Christ B, Wilting J. Endogenous origin of the lymphatics in the avian chorioallantoic membrane. Dev Dyn. 2001; 222:238-51.
52. Wilting J, Papoutsi M, Othman-Hassan K, Rodriguez-Niedenfuhr M, Prols F, Tomarev SI, Eichmann A. Development of the avian lymphatic system. Microsc Res Tech. 2001; 55:81-91.
53. He L, Papoutsi M, Huang R, Tomarev SI, Christ B, Kurz H, Wilting J. Three different fates of cells migrating from somites into the limb bud. Anat Embryol (Berl). 2003; 207:29-34.
54. Oliver G, Sosa-Pineda B, Geisendorf S, Spana EP, Doe CQ, Gruss P. Prox 1, a prospero-related homeobox gene expressed during mouse development. Mech Dev. 1993; 44:3-16.
55. Tomarev SI, Sundin O, Banerjee-Basu S, Duncan MK, Yang JM, Piatigorsky J. Chicken homeobox gene Prox 1 related to Drosophila prospero is expressed in the developing lens and retina. Dev Dyn. 1996; 206:354-67.
56. Schaefer JJ, Oliver G, Henry JJ. Conservation of gene expression during embryonic lens formation and cornea-lens transdifferentiation in Xenopus laevis. Dev Dyn. 1999; 215:308-18.
57. Oliver G, Detmar M. The rediscovery of the lymphatic system: old and new insights into the development and biological function of the lymphatic vasculature. Genes Dev. 2002; 16:773-83.
58. Petrova T, Makinen T, Makela T, Saarela J, Virtanen I, Ferrell R, Finegold D, Kerjaschki D, Yla-Herttuala S, Alitalo K. Lymphatic endothelial reprogramming of vascular endothelial cells by the Prox-1 homeobox transcription factor. EMBO JOURNAL. 2002; 21:4593-4599.
59. Hong Y, Harvey N, Noh Y, Schacht V, Hirakawa S, Detmar M, Oliver G. Prox1 is a master control gene in the program specifying lymphatic endothelial cell fate. Developmental Dynamics. 2002; 225:351-357.
60. Kriehuber E, Breiteneder-Geleff S, Groeger M, Soleiman A, Schoppmann SF, Stingl G, Kerjaschki D, Maurer D. Isolation and characterization of dermal lymphatic and blood endothelial cells reveal stable and functionally specialized cell lineages. J Exp Med. 2001; 194:797-808.
61. Makinen T, Veikkola T, Mustjoki S, Karpanen T, Catimel B, Nice EC, Wise L, Mercer A, Kowalski H, Kerjaschki D, Stacker SA, Achen MG, Alitalo K. Isolated lymphatic endothelial cells transduce growth, survival and migratory signals via the VEGF-C/D receptor VEGFR-3. Embo J. 2001; 20:4762-73.

62. Podgrabinska S, Braun P, Velasco P, Kloos B, Pepper M, Jackson D, M S. Molecular characterization of lymphatic endothelial cells. Proc Natl Acad Sci U S A. 2002; 99:16069-74.
63. Hirakawa S, Hong YK, Harvey N, Schacht V, Matsuda K, Libermann T, Detmar M. Identification of vascular lineage-specific genes by transcriptional profiling of isolated blood vascular and lymphatic endothelial cells. Am J Pathol. 2003; 162:575-86.
64. Huang XZ, Wu JF, Ferrando R, Lee JH, Wang YL, Farese RV, Jr., Sheppard D. Fatal bilateral chylothorax in mice lacking the integrin alpha9beta1. Mol Cell Biol. 2000; 20:5208-15.
65. Wang JF, Zhang XF, Groopman JE. Stimulation of beta 1 integrin induces tyrosine phosphorylation of vascular endothelial growth factor receptor-3 and modulates cell migration. J Biol Chem. 2001; 276:41950-7.
66. Hanahan D, Folkman J. Patterns and emerging mechanisms of the angiogenic switch during tumorigenesis. Cell. 1996; 86:353-64.
67. Kerbel RS, Viloria-Petit A, Okada F, Rak J. Establishing a link between oncogenes and tumor angiogenesis. Mol Med. 1998; 4:286-95.
68. Houck KA, Ferrara N, Winer J, Cachianes G, Li B, Leung DW. The vascular endothelial growth factor family: identification of a fourth molecular species and characterization of alternative splicing of RNA. Mol Endocrinol. 1991; 5:1806-14.
69. Tischer E, Mitchell R, Hartman T, Silva M, Gospodarowicz D, Fiddes JC, Abraham JA. The human gene for vascular endothelial growth factor. Multiple protein forms are encoded through alternative exon splicing. J Biol Chem. 1991; 266:11947-54.
70. Poltorak Z, Cohen T, Neufeld G. The VEGF splice variants: properties, receptors, and usage for the treatment of ischemic diseases. Herz. 2000; 25:126-9.
71. Jingjing L, Xue Y, Agarwal N, Roque RS. Human Muller cells express VEGF183, a novel spliced variant of vascular endothelial growth factor. Invest Ophthalmol Vis Sci. 1999; 40:752-9.
72. Soker S, Takashima S, Miao HQ, Neufeld G, Klagsbrun M. Neuropilin-1 is expressed by endothelial and tumor cells as an isoform-specific receptor for vascular endothelial growth factor. Cell. 1998; 92:735-45.
73. Leung DW, Cachianes G, Kuang WJ, Goeddel DV, Ferrara N. Vascular endothelial growth factor is a secreted angiogenic mitogen. Science. 1989; 246:1306-9.
74. Gospodarowicz D, Abraham JA, Schilling J. Isolation and characterization of a vascular endothelial cell mitogen produced by pituitary-derived folliculo stellate cells. Proc Natl Acad Sci U S A. 1989; 86:7311-5.
75. Keck PJ, Hauser SD, Krivi G, Sanzo K, Warren T, Feder J, Connolly DT. Vascular permeability factor, an endothelial cell mitogen related to PDGF. Science. 1989; 246:1309-12.
76. Connolly DT, Heuvelman DM, Nelson R, Olander JV, Eppley BL, Delfino JJ, Siegel NR, Leimgruber RM, Feder J. Tumor vascular permeability factor stimulates endothelial cell growth and angiogenesis. J Clin Invest. 1989; 84:1470-8.
77. Ferrara N, Henzel WJ. Pituitary follicular cells secrete a novel heparin-binding growth factor specific for vascular endothelial cells. Biochem Biophys Res Commun. 1989; 161:851-8.
78. Plouet J, Schilling J, Gospodarowicz D. Isolation and characterization of a newly identified endothelial cell mitogen produced by AtT-20 cells. Embo J. 1989; 8:3801-6.
79. Conn G, Bayne ML, Soderman DD, Kwok PW, Sullivan KA, Palisi TM, Hope DA, Thomas KA. Amino acid and cDNA sequences of a vascular endothelial cell mitogen that is homologous to platelet-derived growth factor. Proc Natl Acad Sci U S A. 1990; 87:2628-32.

80. Ferrara N, Davis-Smyth T. The biology of vascular endothelial growth factor. Endocr Rev. 1997; 18:4-25.
81. Weinstein BM. What guides early embryonic blood vessel formation? Dev Dyn. 1999; 215:2-11.
82. Carmeliet P, Ferreira V, Breier G, Pollefeyt S, Kieckens L, Gertsenstein M, Fahrig M, Vandenhoeck A, Harpal K, Eberhardt C, Declercq C, Pawling J, Moons L, Collen D, Risau W, Nagy A. Abnormal blood vessel development and lethality in embryos lacking a single VEGF allele. Nature. 1996; 380:435-9.
83. Ferrara N, Carver-Moore K, Chen H, Dowd M, Lu L, O'Shea KS, Powell-Braxton L, Hillan KJ, Moore MW. Heterozygous embryonic lethality induced by targeted inactivation of the VEGF gene. Nature. 1996; 380:439-42.
84. Alon T, Hemo I, Itin A, Pe'er J, Stone J, Keshet E. Vascular endothelial growth factor acts as a survival factor for newly formed retinal vessels and has implications for retinopathy of prematurity. Nat Med. 1995; 1:1024-8.
85. Yuan F, Chen Y, Dellian M, Safabakhsh N, Ferrara N, Jain RK. Time-dependent vascular regression and permeability changes in established human tumor xenografts induced by an anti-vascular endothelial growth factor/vascular permeability factor antibody. Proc Natl Acad Sci U S A. 1996; 93:14765-70.
86. Benjamin LE, Keshet E. Conditional switching of vascular endothelial growth factor (VEGF) expression in tumors: induction of endothelial cell shedding and regression of hemangioblastoma-like vessels by VEGF withdrawal. Proc Natl Acad Sci U S A. 1997; 94:8761-6.
87. Gerber HP, McMurtrey A, Kowalski J, Yan M, Keyt BA, Dixit V, Ferrara N. Vascular endothelial growth factor regulates endothelial cell survival through the phosphatidylinositol 3'-kinase/Akt signal transduction pathway. Requirement for Flk-1/KDR activation. J Biol Chem. 1998; 273:30336-43.
88. Benjamin LE, Hemo I, Keshet E. A plasticity window for blood vessel remodelling is defined by pericyte coverage of the preformed endothelial network and is regulated by PDGF-B and VEGF. Development. 1998; 125:1591-8.
89. Senger DR, Galli SJ, Dvorak AM, Perruzzi CA, Harvey VS, Dvorak HF. Tumor cells secrete a vascular permeability factor that promotes accumulation of ascites fluid. Science. 1983; 219:983-5.
90. Bruce JN, Criscuolo GR, Merrill MJ, Moquin RR, Blacklock JB, Oldfield EH. Vascular permeability induced by protein product of malignant brain tumors: inhibition by dexamethasone. J Neurosurg. 1987; 67:880-4.
91. Dvorak HF, Detmar M, Claffey KP, Nagy JA, van de Water L, Senger DR. Vascular permeability factor/vascular endothelial growth factor: an important mediator of angiogenesis in malignancy and inflammation. Int Arch Allergy Immunol. 1995; 107:233-5.
92. Olofsson B, Pajusola K, Kaipainen A, von Euler G, Joukov V, Saksela O, Orpana A, Pettersson RF, Alitalo K, Eriksson U. Vascular endothelial growth factor B, a novel growth factor for endothelial cells. Proc Natl Acad Sci U S A. 1996; 93:2576-81.
93. Makinen T, Olofsson B, Karpanen T, Hellman U, Soker S, Klagsbrun M, Eriksson U, Alitalo K. Differential binding of vascular endothelial growth factor B splice and proteolytic isoforms to neuropilin-1. J Biol Chem. 1999; 274:21217-22.
94. Li X, Aase K, Li H, von Euler G, Eriksson U. Isoform-specific expression of VEGF-B in normal tissues and tumors. Growth Factors. 2001; 19:49-59.
95. Carmeliet P, Moons L, Luttun A, Vincenti V, Compernolle V, De Mol M, Wu Y, Bono F, Devy L, Beck H, Scholz D, Acker T, DiPalma T, Dewerchin M, Noel A, Stalmans I, Barra A, Blacher S, Vandendriessche T, Ponten A, Eriksson U, Plate KH, Foidart JM,

Schaper W, Charnock-Jones DS, Hicklin DJ, Herbert JM, Collen D, Persico MG. Synergism between vascular endothelial growth factor and placental growth factor contributes to angiogenesis and plasma extravasation in pathological conditions. Nat Med. 2001; 7:575-83.
96. Carmeliet P, Luttun A. The emerging role of the bone marrow-derived stem cells in (therapeutic) angiogenesis. Thromb Haemost. 2001; 86:289-97.
97. Luttun A, Tjwa M, Moons L, Wu Y, Angelillo-Scherrer A, Liao F, Nagy JA, Hooper A, Priller J, De Klerck B, Compernolle V, Daci E, Bohlen P, Dewerchin M, Herbert JM, Fava R, Matthys P, Carmeliet G, Collen D, Dvorak HF, Hicklin DJ, Carmeliet P. Revascularization of ischemic tissues by PlGF treatment, and inhibition of tumor angiogenesis, arthritis and atherosclerosis by anti-Flt1. Nat Med. 2002; 8:831-40.
98. Hattori K, Heissig B, Wu Y, Dias S, Tejada R, Ferris B, Hicklin DJ, Zhu Z, Bohlen P, Witte L, Hendrikx J, Hackett NR, Crystal RG, Moore MA, Werb Z, Lyden D, Rafii S. Placental growth factor reconstitutes hematopoiesis by recruiting VEGFR1(+) stem cells from bone-marrow microenvironment. Nat Med. 2002; 8:841-9.
99. Maglione D, Guerriero V, Viglietto G, Delli-Bovi P, Persico MG. Isolation of a human placenta cDNA coding for a protein related to the vascular permeability factor. Proc Natl Acad Sci U S A. 1991; 88:9267-71.
100. Park JE, Chen HH, Winer J, Houck KA, Ferrara N. Placenta growth factor. Potentiation of vascular endothelial growth factor bioactivity, in vitro and in vivo, and high affinity binding to Flt-1 but not to Flk-1/KDR. J Biol Chem. 1994; 269:25646-54.
101. Maglione D, Guerriero V, Viglietto G, Ferraro MG, Aprelikova O, Alitalo K, Del Vecchio S, Lei KJ, Chou JY, Persico MG. Two alternative mRNAs coding for the angiogenic factor, placenta growth factor (PlGF), are transcribed from a single gene of chromosome 14. Oncogene. 1993; 8:925-31.
102. Cao Y, Ji WR, Qi P, Rosin A. Placenta growth factor: identification and characterization of a novel isoform generated by RNA alternative splicing. Biochem Biophys Res Commun. 1997; 235:493-8.
103. Migdal M, Huppertz B, Tessler S, Comforti A, Shibuya M, Reich R, Baumann H, Neufeld G. Neuropilin-1 is a placenta growth factor-2 receptor. J Biol Chem. 1998; 273:22272-8.
104. Clauss M, Weich H, Breier G, Knies U, Rockl W, Waltenberger J, Risau W. The vascular endothelial growth factor receptor Flt-1 mediates biological activities. Implications for a functional role of placenta growth factor in monocyte activation and chemotaxis. J Biol Chem. 1996; 271:17629-34.
105. Joukov V, Pajusola K, Kaipainen A, Chilov D, Lahtinen I, Kukk E, Saksela O, Kalkkinen N, Alitalo K. A novel vascular endothelial growth factor, VEGF-C, is a ligand for the Flt4 (VEGFR-3) and KDR (VEGFR-2) receptor tyrosine kinases. Embo J. 1996; 15:1751.
106. Lee J, Gray A, Yuan J, Luoh SM, Avraham H, Wood WI. Vascular endothelial growth factor-related protein: a ligand and specific activator of the tyrosine kinase receptor Flt4. Proc Natl Acad Sci U S A. 1996; 93:1988-92.
107. Orlandini M, Marconcini L, Ferruzzi R, Oliviero S. Identification of a c-fos-induced gene that is related to the platelet-derived growth factor/vascular endothelial growth factor family. Proc Natl Acad Sci U S A. 1996; 93:11675-80.
108. Yamada Y, Nezu J, Shimane M, Hirata Y. Molecular cloning of a novel vascular endothelial growth factor, VEGF-D. Genomics. 1997; 42:483-8.
109. Achen MG, Jeltsch M, Kukk E, Makinen T, Vitali A, Wilks AF, Alitalo K, Stacker SA. Vascular endothelial growth factor D (VEGF-D) is a ligand for the tyrosine kinases

VEGF receptor 2 (Flk1) and VEGF receptor 3 (Flt4). Proc Natl Acad Sci U S A. 1998; 95:548-53.

110. Joukov V, Sorsa T, Kumar V, Jeltsch M, Claesson-Welsh L, Cao Y, Saksela O, Kalkkinen N, Alitalo K. Proteolytic processing regulates receptor specificity and activity of VEGF-C. Embo J. 1997; 16:3898-911.
111. Kukk E, Lymboussaki A, Taira S, Kaipainen A, Jeltsch M, Joukov V, Alitalo K. VEGF-C receptor binding and pattern of expression with VEGFR-3 suggests a role in lymphatic vascular development. Development. 1996; 122:3829-37.
112. Eichmann A, Corbel C, Jaffredo T, Breant C, Joukov V, Kumar V, Alitalo K, le Douarin NM. Avian VEGF-C: cloning, embryonic expression pattern and stimulation of the differentiation of VEGFR2-expressing endothelial cell precursors. Development. 1998; 125:743-52.
113. Karkkainen MJ, Haiko P, Sainio K, Partanen J, Taipale J, Petrova TV, Jeltsch M, Jackson DG, Talikka M, Rauvala H, Betsholtz C, Alitalo K. Vascular endothelial growth factor C is required for sprouting of the first lymphatic vessels from embryonic veins. Nat Immunol. 2004; 5:74-80.
114. Partanen TA, Arola J, Saaristo A, Jussila L, Ora A, Miettinen M, Stacker SA, Achen MG, Alitalo K. VEGF-C and VEGF-D expression in neuroendocrine cells and their receptor, VEGFR-3, in fenestrated blood vessels in human tissues. Faseb J. 2000; 14:2087-96.
115. Ristimaki A, Narko K, Enholm B, Joukov V, Alitalo K. Proinflammatory cytokines regulate expression of the lymphatic endothelial mitogen vascular endothelial growth factor-C. J Biol Chem. 1998; 273:8413-8.
116. Skobe M, Brown LF, Tognazzi K, Ganju RK, Dezube BJ, Alitalo K, Detmar M. Vascular endothelial growth factor-C (VEGF-C) and its receptors KDR and flt-4 are expressed in AIDS-associated Kaposi's sarcoma. J Invest Dermatol. 1999; 113:1047-53.
117. Jeltsch M, Kaipainen A, Joukov V, Meng X, Lakso M, Rauvala H, Swartz M, Fukumura D, Jain RK, Alitalo K. Hyperplasia of lymphatic vessels in VEGF-C transgenic mice. Science. 1997; 276:1423-5.
118. Veikkola T, Jussila L, Makinen T, Karpanen T, Jeltsch M, Petrova TV, Kubo H, Thurston G, McDonald DM, Achen MG, Stacker SA, Alitalo K. Signalling via vascular endothelial growth factor receptor-3 is sufficient for lymphangiogenesis in transgenic mice. Embo J. 2001; 20:1223-31.
119. Makinen T, Jussila L, Veikkola T, Karpanen T, Kettunen MI, Pulkkanen KJ, Kauppinen R, Jackson DG, Kubo H, Nishikawa S, Yla-Herttuala S, Alitalo K. Inhibition of lymphangiogenesis with resulting lymphedema in transgenic mice expressing soluble VEGF receptor-3. Nat Med. 2001; 7:199-205.
120. Cao Y, Linden P, Farnebo J, Cao R, Eriksson A, Kumar V, Qi JH, Claesson-Welsh L, Alitalo K. Vascular endothelial growth factor C induces angiogenesis in vivo. Proc Natl Acad Sci U S A. 1998; 95:14389-94.
121. Witzenbichler B, Maisonpierre PC, Jones P, Yancopoulos GD, Isner JM. Chemotactic properties of angiopoietin-1 and -2, ligands for the endothelial-specific receptor tyrosine kinase Tie2. J Biol Chem. 1998; 273:18514-21.
122. Stacker SA, Stenvers K, Caesar C, Vitali A, Domagala T, Nice E, Roufail S, Simpson RJ, Moritz R, Karpanen T, Alitalo K, Achen MG. Biosynthesis of vascular endothelial growth factor-D involves proteolytic processing which generates non-covalent homodimers. J Biol Chem. 1999; 274:32127-36.
123. Baldwin ME, Roufail S, Halford MM, Alitalo K, Stacker SA, Achen MG. Multiple forms of mouse vascular endothelial growth factor-D are generated by RNA splicing and proteolysis. J Biol Chem. 2001; 276:44307-14.

124. Baldwin ME, Catimel B, Nice EC, Roufail S, Hall NE, Stenvers KL, Karkkainen MJ, Alitalo K, Stacker SA, Achen MG. The specificity of receptor binding by vascular endothelial growth factor-d is different in mouse and man. J Biol Chem. 2001; 276:19166-71.
125. Avantaggiato V, Orlandini M, Acampora D, Oliviero S, Simeone A. Embryonic expression pattern of the murine figf gene, a growth factor belonging to platelet-derived growth factor/vascular endothelial growth factor family. Mech Dev. 1998; 73:221-4.
126. Farnebo F, Piehl F, Lagercrantz J. Restricted expression pattern of vegf-d in the adult and fetal mouse: high expression in the embryonic lung. Biochem Biophys Res Commun. 1999; 257:891-4.
127. Marconcini L, Marchio S, Morbidelli L, Cartocci E, Albini A, Ziche M, Bussolino F, Oliviero S. c-fos-induced growth factor/vascular endothelial growth factor D induces angiogenesis in vivo and in vitro. Proc Natl Acad Sci U S A. 1999; 96:9671-6.
128. Neufeld G, Cohen T, Gengrinovitch S, Poltorak Z. Vascular endothelial growth factor (VEGF) and its receptors. Faseb J. 1999; 13:9-22.
129. Karkkainen MJ, Ferrell RE, Lawrence EC, Kimak MA, Levinson KL, McTigue MA, Alitalo K, Finegold DN. Missense mutations interfere with VEGFR-3 signalling in primary lymphoedema. Nat Genet. 2000; 25:153-9.
130. Shibuya M, Yamaguchi S, Yamane A, Ikeda T, Tojo A, Matsushime H, Sato M. Nucleotide sequence and expression of a novel human receptor-type tyrosine kinase gene (flt) closely related to the fms family. Oncogene. 1990; 5:519-24.
131. de Vries C, Escobedo JA, Ueno H, Houck K, Ferrara N, Williams LT. The fms-like tyrosine kinase, a receptor for vascular endothelial growth factor. Science. 1992; 255:989-91.
132. Kendall RL, Thomas KA. Inhibition of vascular endothelial cell growth factor activity by an endogenously encoded soluble receptor. Proc Natl Acad Sci U S A. 1993; 90:10705-9.
133. Gluzman-Poltorak Z, Cohen T, Shibuya M, Neufeld G. Vascular endothelial growth factor receptor-1 and neuropilin-2 form complexes. J Biol Chem. 2001; 276:18688-94.
134. Fong GH, Rossant J, Gertsenstein M, Breitman ML. Role of the Flt-1 receptor tyrosine kinase in regulating the assembly of vascular endothelium. Nature. 1995; 376:66-70.
135. Fong GH, Zhang L, Bryce DM, Peng J. Increased hemangioblast commitment, not vascular disorganization, is the primary defect in flt-1 knock-out mice. Development. 1999; 126:3015-25.
136. Hiratsuka S, Minowa O, Kuno J, Noda T, Shibuya M. Flt-1 lacking the tyrosine kinase domain is sufficient for normal development and angiogenesis in mice. Proc Natl Acad Sci U S A. 1998; 95:9349-54.
137. Hong YK, Lange-Asschenfeldt B, Velasco P, Hirakawa S, Kunstfeld R, Brown LF, Bohlen P, Senger DR, Detmar M. VEGF-A promotes tissue repair-associated lymphatic vessel formation via VEGFR-2 and the alpha1beta1 and alpha2beta1 integrins. Faseb J. 2004; 18:1111-3.
138. Shalaby F, Rossant J, Yamaguchi TP, Gertsenstein M, Wu XF, Breitman ML, Schuh AC. Failure of blood-island formation and vasculogenesis in Flk-1-deficient mice. Nature. 1995; 376:62-6.
139. Shalaby F, Ho J, Stanford WL, Fischer KD, Schuh AC, Schwartz L, Bernstein A, Rossant J. A requirement for Flk1 in primitive and definitive hematopoiesis and vasculogenesis. Cell. 1997; 89:981-90.
140. Ziegler BL, Valtieri M, Porada GA, De Maria R, Muller R, Masella B, Gabbianelli M, Casella I, Pelosi E, Bock T, Zanjani ED, Peschle C. KDR receptor: a key marker defining hematopoietic stem cells. Science. 1999; 285:1553-8.

141. Saaristo A, Veikkola T, Enholm B, Hytonen M, Arola J, Pajusola K, Turunen P, Jeltsch M, Karkkainen MJ, Kerjaschki D, Bueler H, Yla-Herttuala S, Alitalo K. Adenoviral VEGF-C overexpression induces blood vessel enlargement, tortuosity, and leakiness but no sprouting angiogenesis in the skin or mucous membranes. Faseb J. 2002; 16:1041-9.
142. Matsumoto T, Claesson-Welsh L. VEGF receptor signal transduction. Sci STKE. 2001; 2001:RE21.
143. Eliceiri BP, Paul R, Schwartzberg PL, Hood JD, Leng J, Cheresh DA. Selective requirement for Src kinases during VEGF-induced angiogenesis and vascular permeability. Mol Cell. 1999; 4:915-24.
144. Paul R, Zhang ZG, Eliceiri BP, Jiang Q, Boccia AD, Zhang RL, Chopp M, Cheresh DA. Src deficiency or blockade of Src activity in mice provides cerebral protection following stroke. Nat Med. 2001; 7:222-7.
145. Nagy JA, Vasile E, Feng D, Sundberg C, Brown LF, Detmar MJ, Lawitts JA, Benjamin L, Tan X, Manseau EJ, Dvorak AM, Dvorak HF. Vascular permeability factor/vascular endothelial growth factor induces lymphangiogenesis as well as angiogenesis. J Exp Med. 2002; 196:1497-506.
146. Enholm B, Karpanen T, Jeltsch M, Kubo H, Stenback F, Prevo R, Jackson DG, Yla-Herttuala S, Alitalo K. Adenoviral expression of vascular endothelial growth factor-C induces lymphangiogenesis in the skin. Circ Res. 2001; 88:623-9.
147. Byzova T, Goldman C, Jankau J, Chen J, Cabrera G, Achen M, Stacker S, Carnevale K, Siemionow M, Deitcher S, DiCorleto P. Adenovirus encoding vascular endothelial growth factor-D induces tissue-specific vascular patterns in vivo. Blood. 2002; 99:4434-42.
148. Kunstfeld R, Hirakawa S, Hong YK, Schacht V, Lange-Asschenfeldt B, Velasco P, Lin C, Fiebiger E, Wei X, Wu Y, Hicklin D, Bohlen P, Detmar M. Induction of cutaneous delayed-type hypersensitivity reactions in VEGF-A transgenic mice results in chronic skin inflammation associated with persistent lymphatic hyperplasia. Blood. 2004; 104:1048-57.
149. Kaipainen A, Korhonen J, Pajusola K, Aprelikova O, Persico MG, Terman BI, Alitalo K. The related FLT4, FLT1, and KDR receptor tyrosine kinases show distinct expression patterns in human fetal endothelial cells. J Exp Med. 1993; 178:2077-88.
150. Pajusola K, Aprelikova O, Armstrong E, Morris S, Alitalo K. Two human FLT4 receptor tyrosine kinase isoforms with distinct carboxy terminal tails are produced by alternative processing of primary transcripts. Oncogene. 1993; 8:2931-7.
151. Pajusola K, Aprelikova O, Korhonen J, Kaipainen A, Pertovaara L, Alitalo R, Alitalo K. FLT4 receptor tyrosine kinase contains seven immunoglobulin-like loops and is expressed in multiple human tissues and cell lines. Cancer Res. 1992; 52:5738-43.
152. Aprelikova O, Pajusola K, Partanen J, Armstrong E, Alitalo R, Bailey SK, McMahon J, Wasmuth J, Huebner K, Alitalo K. FLT4, a novel class III receptor tyrosine kinase in chromosome 5q33-qter. Cancer Res. 1992; 52:746-8.
153. Galland F, Karamysheva A, Pebusque MJ, Borg JP, Rottapel R, Dubreuil P, Rosnet O, Birnbaum D. The FLT4 gene encodes a transmembrane tyrosine kinase related to the vascular endothelial growth factor receptor. Oncogene. 1993; 8:1233-40.
154. Ferrell R. Research perspectives in inherited lymphatic disease. Ann N Y Acad Sci. 2002; 979:39-51.
155. Joukov V, Kumar V, Sorsa T, Arighi E, Weich H, Saksela O, Alitalo K. A recombinant mutant vascular endothelial growth factor-C that has lost vascular endothelial growth factor receptor-2 binding, activation, and vascular permeability activities. J Biol Chem. 1998; 273:6599-602.

156. Yancopoulos GD, Davis S, Gale NW, Rudge JS, Wiegand SJ, Holash J. Vascular-specific growth factors and blood vessel formation. Nature. 2000; 407:242-8.
157. Jones N, Iljin K, Dumont DJ, Alitalo K. Tie receptors: new modulators of angiogenic and lymphangiogenic responses. Nat Rev Mol Cell Biol. 2001; 2:257-67.
158. Loughna S, Sato TN. Angiopoietin and Tie signaling pathways in vascular development. Matrix Biol. 2001; 20:319-25.
159. Iljin K, Petrova TV, Veikkola T, Kumar V, Poutanen M, Alitalo K. A fluorescent Tie1 reporter allows monitoring of vascular development and endothelial cell isolation from transgenic mouse embryos. Faseb J. 2002; 16:1764-74.
160. Sato TN, Tozawa Y, Deutsch U, Wolburg-Buchholz K, Fujiwara Y, Gendron-Maguire M, Gridley T, Wolburg H, Risau W, Qin Y. Distinct roles of the receptor tyrosine kinases Tie-1 and Tie-2 in blood vessel formation. Nature. 1995; 376:70-4.
161. Puri MC, Rossant J, Alitalo K, Bernstein A, Partanen J. The receptor tyrosine kinase TIE is required for integrity and survival of vascular endothelial cells. Embo J. 1995; 14:5884-91.
162. Dumont DJ, Gradwohl G, Fong GH, Puri MC, Gertsenstein M, Auerbach A, Breitman ML. Dominant-negative and targeted null mutations in the endothelial receptor tyrosine kinase, tek, reveal a critical role in vasculogenesis of the embryo. Genes Dev. 1994; 8:1897-909.
163. Davis S, Aldrich TH, Jones PF, Acheson A, Compton DL, Jain V, Ryan TE, Bruno J, Radziejewski C, Maisonpierre PC, Yancopoulos GD. Isolation of angiopoietin-1, a ligand for the TIE2 receptor, by secretion-trap expression cloning. Cell. 1996; 87: 1161-9.
164. Maisonpierre PC, Suri C, Jones PF, Bartunkova S, Wiegand SJ, Radziejewski C, Compton D, McClain J, Aldrich TH, Papadopoulos N, Daly TJ, Davis S, Sato TN, Yancopoulos GD. Angiopoietin-2, a natural antagonist for Tie2 that disrupts in vivo angiogenesis. Science. 1997; 277:55-60.
165. Kim I, Kwak HJ, Ahn JE, So JN, Liu M, Koh KN, Koh GY. Molecular cloning and characterization of a novel angiopoietin family protein, angiopoietin-3. FEBS Lett. 1999; 443:353-6.
166. Valenzuela DM, Griffiths JA, Rojas J, Aldrich TH, Jones PF, Zhou H, McClain J, Copeland NG, Gilbert DJ, Jenkins NA, Huang T, Papadopoulos N, Maisonpierre PC, Davis S, Yancopoulos GD. Angiopoietins 3 and 4: diverging gene counterparts in mice and humans. Proc Natl Acad Sci U S A. 1999; 96:1904-9.
167. Suri C, Jones PF, Patan S, Bartunkova S, Maisonpierre PC, Davis S, Sato TN, Yancopoulos GD. Requisite role of angiopoietin-1, a ligand for the TIE2 receptor, during embryonic angiogenesis. Cell. 1996; 87:1171-80.
168. Thurston G, Rudge JS, Ioffe E, Zhou H, Ross L, Croll SD, Glazer N, Holash J, McDonald DM, Yancopoulos GD. Angiopoietin-1 protects the adult vasculature against plasma leakage. Nat Med. 2000; 6:460-3.
169. Mandriota SJ, Pepper MS. Regulation of angiopoietin-2 mRNA levels in bovine microvascular endothelial cells by cytokines and hypoxia. Circ Res. 1998; 83:852-9.
170. Mandriota SJ, Pyke C, Di Sanza C, Quinodoz P, Pittet B, Pepper MS. Hypoxia-inducible angiopoietin-2 expression is mimicked by iodonium compounds and occurs in the rat brain and skin in response to systemic hypoxia and tissue ischemia. Am J Pathol. 2000; 156:2077-89.
171. Holash J, Maisonpierre PC, Compton D, Boland P, Alexander CR, Zagzag D, Yancopoulos GD, Wiegand SJ. Vessel cooption, regression, and growth in tumors mediated by angiopoietins and VEGF. Science. 1999; 284:1994-8.

172. Holash J, Wiegand SJ, Yancopoulos GD. New model of tumor angiogenesis: dynamic balance between vessel regression and growth mediated by angiopoietins and VEGF. Oncogene. 1999; 18:5356-62.
173. Gale NW, Thurston G, Hackett SF, Renard R, Wang Q, McClain J, Martin C, Witte C, Witte MH, Jackson D, Suri C, Campochiaro PA, Wiegand SJ, Yancopoulos GD. Angiopoietin-2 is required for postnatal angiogenesis and lymphatic patterning, and only the latter role is rescued by Angiopoietin-1. Dev Cell. 2002; 3:411-23.
174. Banerji S, Ni J, Wang SX, Clasper S, Su J, Tammi R, Jones M, Jackson DG. LYVE-1, a new homologue of the CD44 glycoprotein, is a lymph-specific receptor for hyaluronan. J Cell Biol. 1999; 144:789-801.
175. Jackson DG, Prevo R, Clasper S, Banerji S. LYVE-1, the lymphatic system and tumor lymphangiogenesis. Trends Immunol. 2001; 22:317-21.
176. Jackson DG. The lymphatics revisited: new perspectives from the hyaluronan receptor LYVE-1. Trends Cardiovasc Med. 2003; 13:1-7.
177. Wetterwald A, Hoffstetter W, Cecchini MG, Lanske B, Wagner C, Fleisch H, Atkinson M. Characterization and cloning of the E11 antigen, a marker expressed by rat osteoblasts and osteocytes. Bone. 1996; 18:125-32.
178. Schacht V, Ramirez MI, Hong YK, Hirakawa S, Feng D, Harvey N, Williams M, Dvorak AM, Dvorak HF, Oliver G, Detmar M. T1alpha/podoplanin deficiency disrupts normal lymphatic vasculature formation and causes lymphedema. Embo J. 2003; 22:3546-56.
179. Ramirez MI, Millien G, Hinds A, Cao Y, Seldin DC, Williams MC. T1alpha, a lung type I cell differentiation gene, is required for normal lung cell proliferation and alveolus formation at birth. Dev Biol. 2003; 256:61-72.
180. Rishi AK, Joyce-Brady M, Fisher J, Dobbs LG, Floros J, VanderSpek J, Brody JS, Williams MC. Cloning, characterization, and development expression of a rat lung alveolar type I cell gene in embryonic endodermal and neural derivatives. Dev Biol. 1995; 167:294-306.
181. Hong YK, Detmar M. Prox1, master regulator of the lymphatic vasculature phenotype. Cell Tissue Res. 2003; 314:85-92.
182. Neufeld G, Cohen T, Shraga N, Lange T, Kessler O, Herzog Y. The neuropilins: multifunctional semaphorin and VEGF receptors that modulate axon guidance and angiogenesis. Trends Cardiovasc Med. 2002; 12:13-9.
183. Giraudo E, Primo L, Audero E, Gerber HP, Koolwijk P, Soker S, Klagsbrun M, Ferrara N, Bussolino F. Tumor necrosis factor-alpha regulates expression of vascular endothelial growth factor receptor-2 and of its co-receptor neuropilin-1 in human vascular endothelial cells. J Biol Chem. 1998; 273: 22128-35.
184. Arruda VR, Hagstrom JN, Deitch J, Heiman-Patterson T, Camire RM, Chu K, Fields PA, Herzog RW, Couto LB, Larson PJ, High KA. Posttranslational modifications of recombinant myotube-synthesized human factor IX. Blood. 2001; 97:130-8.
185. Karkkainen MJ, Saaristo A, Jussila L, Karila KA, Lawrence EC, Pajusola K, Bueler H, Eichmann A, Kauppinen R, Kettunen MI, Yla-Herttuala S, Finegold DN, Ferrell RE, Alitalo K. A model for gene therapy of human hereditary lymphedema. Proc Natl Acad Sci U S A. 2001; 98:12677-82.
186. Paavonen K, Puolakkainen P, Jussila L, Jahkola T, Alitalo K. Vascular endothelial growth factor receptor-3 in lymphangiogenesis in wound healing. Am J Pathol. 2000; 156:1499-504.
187. Clark E, Clark E. Observations on the new growth of lymphatic vessels as seen in transparent chambers introduced into the rabbit's ear. Am J Anat. 1932; 51:43-87.

188. Rockson S. Primary Lymphedema. In: Ernst C, Stanley J, eds. Current Therapy in Vascular Surgery. Fourth ed. Philadelphia: Mosby; 2000:915-918.
189. Witte M, Way D, Witte C, Bernas MLM, significance and clinical implications. In: Goldberg ID, Rosen EM, editors. Regulation of angiogenesis. Basel, Switzerland: Birkhauser Verlag; 1997. 65-112. Lymphangiogenesis: Mechanisms, significance and clinical implications. In: Goldberg I, Rosen E, eds. Regulation of angiogenesis. Basel, Switzerland: Birkhauser Verlag; 1997:65-112.
190. Karpanen T, Alitalo K. Lymphatic vessels as targets of tumor therapy? J Exp Med. 2001; 194:F37-42.
191. Rockson SG. Lymphedema. Am J Med. 2001; 110:288-95.
192. Szuba A, Rockson SG. Lymphedema: classification, diagnosis and therapy. Vasc Med. 1998; 3:145-56.
193. Shin WS, Szuba A, Rockson SG. Animal models for the study of lymphatic insufficiency. Lymphat Res Biol. 2003; 1:159-69.
194. Piller NB. Lymphoedema, macrophages and benzopyrones. Lymphology. 1980; 13: 109-19.
195. Piller NB. Macrophage and tissue changes in the developmental phases of secondary lymphoedema and during conservative therapy with benzopyrone. Arch Histol Cytol. 1990; 53:209-18.
196. Szuba A, Skobe M, Karkkainen MJ, Shin WS, Beynet DP, Rockson NB, Dakhil N, Spilman S, Goris ML, Strauss HW, Quertermous T, Alitalo K, Rockson SG. Therapeutic lymphangiogenesis with human recombinant VEGF-C. FASEB J. 2002; 16(14): 1985-1987.
197. Velanovich V, Szymanski W. Quality of life of breast cancer patients with lymphedema. Am J Surg. 1999; 177:184-7; discussion 188.
198. Tobin MB, Lacey HJ, Meyer L, Mortimer PS. The psychological morbidity of breast cancer-related arm swelling. Psychological morbidity of lymphoedema. Cancer. 1993; 72:3248-52.
199. Maunsell E, Brisson J, Deschenes L. Arm problems and psychological distress after surgery for breast cancer. Can J Surg. 1993; 36:315-20.
200. Passik S, Newman M, Brennan M, Tunkel R. Predictors of psychological distress, sexual dysfunction and physical functioning among women with upper extremity lymphedema related to breast cancer. Psycho-Oncology. 1995; 4:255-263.
201. Rockson SG. Lymphedema after surgery for cancer: the role of patient support groups in patient therapy. Disease Management and Health Outcomes. 2002; 10:345-7.
202. Rockson SG, Miller LT, Senie R, Brennan MJ, Casley-Smith JR, Foldi E, Foldi M, Gamble GL, Kasseroller RG, Leduc A, Lerner R, Mortimer PS, Norman SA, Plotkin CL, Rinehart-Ayres ME, Walder AL. American Cancer Society Lymphedema Workshop. Workgroup III: Diagnosis and management of lymphedema. Cancer. 1998; 83:2882-5.
203. Rivard A, Isner JM. Angiogenesis and vasculogenesis in treatment of cardiovascular disease. Mol Med. 1998; 4:429-40.
204. Baumgartner I, Pieczek A, Manor O, Blair R, Kearney M, Walsh K, Isner JM. Constitutive expression of phVEGF165 after intramuscular gene transfer promotes collateral vessel development in patients with critical limb ischemia [see comments]. Circulation. 1998; 97:1114-23.
205. Ferrara N, Alitalo K. Clinical applications of angiogenic growth factors and their inhibitors. Nat Med. 1999; 5:1359-64.

206. Ferrell RE, Levinson KL, Esman JH, Kimak MA, Lawrence EC, Barmada MM, Finegold DN. Hereditary lymphedema: evidence for linkage and genetic heterogeneity. Hum Mol Genet. 1998; 7:2073-8.
207. Evans AL, Brice G, Sotirova V, Mortimer P, Beninson J, Burnand K, Rosbotham J, Child A, Sarfarazi M. Mapping of primary congenital lymphedema to the 5q35.3 region. Am J Hum Genet. 1999; 64:547-55.
208. Witte MH, Erickson R, Bernas M, Andrade M, Reiser F, Conlon W, Hoyme HE, Witte CL. Phenotypic and genotypic heterogeneity in familial Milroy lymphedema. Lymphology. 1998; 31:145-55.
209. Karkkainen MJ, Petrova TV. Vascular endothelial growth factor receptors in the regulation of angiogenesis and lymphangiogenesis. Oncogene. 2000; 19:5598-605.
210. Karkkainen MJ, Jussila L, Ferrell RE, Finegold DN, Alitalo K. Molecular regulation of lymphangiogenesis and targets for tissue oedema. Trends Mol Med. 2001; 7:18-22.
211. Barber JC, Temple IK, Campbell PL, Collinson MN, Campbell CM, Renshaw RM, Dennis NR. Unbalanced translocation in a mother and her son in one of two 5;10 translocation families. Am J Med Genet. 1996; 62:84-90.
212. Groen SE, Drewes JG, de Boer EG, Hoovers JM, Hennekam RC. Repeated unbalanced offspring due to a familial translocation involving chromosomes 5 and 6. Am J Med Genet. 1998; 80:448-53.
213. Fang J, Dagenais SL, Erickson RP, Arlt MF, Glynn MW, Gorski JL, Seaver LH, Glover TW. Mutations in FOXC2 (MFH-1), a forkhead family transcription factor, are responsible for the hereditary lymphedema-distichiasis syndrome [In Process Citation]. Am J Hum Genet. 2000; 67:1382-8.
214. Dale RF. Primary lymphoedema when found with distichiasis is of the type defined as bilateral hyperplasia by lymphography. J Med Genet. 1987; 24:170-1.
215. Falls HF, Kertesz ED. A New Syndrome Combining Pterygium Colli with Developmental Anomalies of the Eyelids and Lymphatics of the Lower Extremities. Trans Am Ophthalmol Soc. 1964; 62:248-75.
216. Chynn KY. Congenital spinal extradural cyst in two siblings. Am J Roentgenol Radium Ther Nucl Med. 1967; 101:204-15.
217. Pap Z, Biro T, Szabo L, Papp Z. Syndrome of lymphoedema and distichiasis. Hum Genet. 1980; 53:309-10.
218. Corbett CR, Dale RF, Coltart DJ, Kinmonth JB. Congenital heart disease in patients with primary lymphedemas. Lymphology. 1982; 15:85-90.
219. Goldstein S, Qazi QH, Fitzgerald J, Goldstein J, Friedman AP, Sawyer P. Distichiasis, congenital heart defects and mixed peripheral vascular anomalies. Am J Med Genet. 1985; 20:283-94.
220. Bartley GB, Jackson IT. Distichiasis and cleft palate. Plast Reconstr Surg. 1989; 84:129-32.
221. Mangion J, Rahman N, Mansour S, Brice G, Rosbotham J, Child AH, Murday VA, Mortimer PS, Barfoot R, Sigurdsson A, Edkins S, Sarfarazi M, Burnand K, Evans AL, Nunan TO, Stratton MR, Jeffery S. A gene for lymphedema-distichiasis maps to 16q24.3. Am J Hum Genet. 1999; 65:427-32.
222. Bell R, Brice G, Child AH, Murday VA, Mansour S, Sandy CJ, Collin JR, Mortimer P, Callen DF, Burnand K. Reduction of the genetic interval for lyphoedema-distichiasis to below 2 Mb. J Med Genet. 2000; 37:725.
223. Finegold DN, Kimak MA, Lawrence EC, Levinson KL, Cherniske EM, Pober BR, Dunlap JW, Ferrell RE. Truncating mutations in FOXC2 cause multiple lymphedema syndromes. Hum Mol Genet. 2001; 10:1185-9.

224. Erickson RP, Dagenais SL, Caulder MS, Downs CA, Herman G, Jones MC, Kerstjens-Frederikse WS, Lidral AC, McDonald M, Nelson CC, Witte M, Glover TW. Clinical heterogeneity in lymphoedema-distichiasis with FOXC2 truncating mutations. J Med Genet. 2001; 38:761-6.
225. Bell R, Brice G, Child AH, Murday VA, Mansour S, Sandy CJ, Collin JR, Brady AF, Callen DF, Burnand K, Mortimer P, Jeffery S. Analysis of lymphoedema-distichiasis families for FOXC2 mutations reveals small insertions and deletions throughout the gene. Hum Genet. 2001; 108:546-51.
226. Weigel D, Jurgens G, Kuttner F, Seifert E, Jackle H. The homeotic gene fork head encodes a nuclear protein and is expressed in the terminal regions of the Drosophila embryo. Cell. 1989; 57:645-58.
227. Winnier GE, Hargett L, Hogan BL. The winged helix transcription factor MFH1 is required for proliferation and patterning of paraxial mesoderm in the mouse embryo. Genes Dev. 1997; 11:926-40.
228. Iida K, Koseki H, Kakinuma H, Kato N, Mizutani-Koseki Y, Ohuchi H, Yoshioka H, Noji S, Kawamura K, Kataoka Y, Ueno F, Taniguchi M, Yoshida N, Sugiyama T, Miura N. Essential roles of the winged helix transcription factor MFH-1 in aortic arch patterning and skeletogenesis. Development. 1997; 124:4627-38.
229. Smith RS, Zabaleta A, Kume T, Savinova OV, Kidson SH, Martin JE, Nishimura DY, Alward WL, Hogan BL, John SW. Haploinsufficiency of the transcription factors FOXC1 and FOXC2 results in aberrant ocular development. Hum Mol Genet. 2000; 9:1021-32.
230. Olszewski W, Machowski Z, Sokolowski J, Nielubowicz J. Experimental lymphedema in dogs. J Cardiovasc Surg (Torino). 1968; 9:178-83.
231. Rockson SG. Preclinical models of lymphatic disease: the potential for growth factor and gene therapy. Ann N Y Acad Sci. 2002; 979:64-75; discussion 76-9.
232. Wang GY, Zhong SZ. A model of experimental lymphedema in rats' limbs. Microsurgery. 1985; 6:204-10.
233. Slavin SA, Van den Abbeele AD, Losken A, Swartz MA, Jain RK. Return of lymphatic function after flap transfer for acute lymphedema. Ann Surg. 1999; 229:421-7.
234. Swartz MA, Berk DA, Jain RK. Transport in lymphatic capillaries. I. Macroscopic measurements using residence time distribution theory. Am J Physiol. 1996; 270: H324-9.
235. Kriederman B, Myloyde T, Bernas M, Lee-Donaldson L, Preciado S, Lynch M, Stea B, Summers P, Witte C, Witte M. Limb volume reduction after physical treatment by compression and/or massage in a rodent model of peripheral lymphedema. Lymphology. 2002; 35:23-7.
236. Lee-Donaldson L, Witte MH, Bernas M, Witte CL, Way D, Stea B. Refinement of a rodent model of peripheral lymphedema. Lymphology. 1999; 32:111-7.
237. Gray H. Studies of the regeneration of lymphatic vessels. J Anat. 1939;74:309.
238. Goffrini P, Bobbio P. [the Lymph Circulation of the Upper Extremity Following the Radical Operation of Mammary Cancer and Its Relations to the Secondary Edema of the Arm.]. Chirurg. 1964; 35:145-8.
239. Huang GK, Hsin YP. An experimental model for lymphedema in rabbit ear. Microsurgery. 1983; 4:236-42.
240. Blau HM, Banfi A. The well-tempered vessel. Nat Med. 2001; 7:532-4.
241. Yla-Herttuala S. Gene therapy for coronary heart disease. J Intern Med. 2001; 250: 367-8.
242. Isner JM. Myocardial gene therapy. Nature. 2002; 415:234-9.

243. Yla-Herttuala S, Alitalo K. Gene transfer as a tool to induce therapeutic vascular growth. Nat Med. 2003; 9:694-701.
244. Mandriota SJ, Jussila L, Jeltsch M, Compagni A, Baetens D, Prevo R, Banerji S, Huarte J, Montesano R, Jackson DG, Orci L, Alitalo K, Christofori G, Pepper MS. Vascular endothelial growth factor-C-mediated lymphangiogenesis promotes tumour metastasis. Embo J. 2001; 20:672-82.
245. Skobe M, Hawighorst T, Jackson DG, Prevo R, Janes L, Velasco P, Riccardi L, Alitalo K, Claffey K, Detmar M. Induction of tumor lymphangiogenesis by VEGF-C promotes breast cancer metastasis. Nat Med. 2001; 7:192-8.
246. Stacker SA, Caesar C, Baldwin ME, Thornton GE, Williams RA, Prevo R, Jackson DG, Nishikawa S, Kubo H, Achen MG. VEGF-D promotes the metastatic spread of tumor cells via the lymphatics. Nat Med. 2001; 7:186-91.
247. Karpanen T, Egeblad M, Karkkainen MJ, Kubo H, Yla-Herttuala S, Jaattela M, Alitalo K. Vascular endothelial growth factor C promotes tumor lymphangiogenesis and intralymphatic tumor growth. Cancer Res. 2001; 61:1786-90.
248. He Y, Kozaki K, Karpanen T, Koshikawa K, Yla-Herttuala S, Takahashi T, Alitalo K. Suppression of tumor lymphangiogenesis and lymph node metastasis by blocking vascular endothelial growth factor receptor 3 signaling. J Natl Cancer Inst. 2002; 94:819-25.
249. Mattila MM, Ruohola JK, Karpanen T, Jackson DG, Alitalo K, Harkonen PL. VEGF-C induced lymphangiogenesis is associated with lymph node metastasis in orthotopic MCF-7 tumors. Int J Cancer. 2002; 98:946-51.
250. Saaristo A, Karkkainen MJ, Alitalo K. Insights into the molecular pathogenesis and targeted treatment of lymphedema. Ann N Y Acad Sci. 2002; 979:94-110.
251. Daly TM, Ohlemiller KK, Roberts MS, Vogler CA, Sands MS. Prevention of systemic clinical disease in MPS VII mice following AAV-mediated neonatal gene transfer. Gene Ther. 2001; 8:1291-8.
252. Monahan PE, Samulski RJ. Adeno-associated virus vectors for gene therapy: more pros than cons? Mol Med Today. 2000; 6:433-40.
253. Saaristo A, Veikkola T, Tammela T, Enholm B, Karkkainen MJ, Pajusola K, Bueler H, Yla-Herttuala S, Alitalo K. Lymphangiogenic gene therapy with minimal blood vascular side effects. J Exp Med. 2002; 196:719-30.
254. Carmeliet P. VEGF gene therapy: stimulating angiogenesis or angioma-genesis? Nat Med. 2000; 6:1102-3.
255. Epstein SE, Kornowski R, Fuchs S, Dvorak HF. Angiogenesis therapy: amidst the hype, the neglected potential for serious side effects. Circulation. 2001; 104:115-9.
256. Thurston G, Suri C, Smith K, McClain J, Sato TN, Yancopoulos GD, McDonald DM. Leakage-resistant blood vessels in mice transgenically overexpressing angiopoietin-1. Science. 1999; 286:2511-4.
257. Witzenbichler B, Asahara T, Murohara T, Silver M, Spyridopoulos I, Magner M, Principe N, Kearney M, Hu JS, Isner JM. Vascular endothelial growth factor-C (VEGF-C/VEGF-2) promotes angiogenesis in the setting of tissue ischemia. Am J Pathol. 1998; 153:381-94.
258. Suri C, McClain J, Thurston G, McDonald DM, Zhou H, Oldmixon EH, Sato TN, Yancopoulos GD. Increased vascularization in mice overexpressing angiopoietin-1. Science. 1998; 282:468-71.
259. Liotta LA, Stetler-Stevenson WG, Steeg PS. Cancer invasion and metastasis: positive and negative regulatory elements. Cancer Invest. 1991; 9:543-51.
260. Cassella M, Skobe M. Lymphatic vessel activation in cancer. Ann N Y Acad Sci. 2002; 979:120-30.

261. Zetter BR. Adhesion molecules in tumor metastasis. Semin Cancer Biol. 1993; 4: 219-29.
262. Padera TP, Kadambi A, di Tomaso E, Carreira CM, Brown EB, Boucher Y, Choi NC, Mathisen D, Wain J, Mark EJ, Munn LL, Jain RK. Lymphatic metastasis in the absence of functional intratumor lymphatics. Science. 2002; 296:1883-6.
263. Skobe M, Hamberg LM, Hawighorst T, Schirner M, Wolf GL, Alitalo K, Detmar M. Concurrent induction of lymphangiogenesis, angiogenesis, and macrophage recruitment by vascular endothelial growth factor-C in melanoma. Am J Pathol. 2001; 159:893-903.
264. Birner P, Schindl M, Obermair A, Plank C, Breitenecker G, Kowalski H, Oberhuber G. Lymphatic microvessel density in epithelial ovarian cancer: its impact on prognosis. Anticancer Res. 2000; 20:2981-5.
265. Birner P, Schindl M, Obermair A, Breitenecker G, Kowalski H, Oberhuber G. Lymphatic microvessel density as a novel prognostic factor in early-stage invasive cervical cancer. Int J Cancer. 2001; 95:29-33.
266. de Waal RM, van Altena MC, Erhard H, Weidle UH, Nooijen PT, Ruiter DJ. Lack of lymphangiogenesis in human primary cutaneous melanoma. Consequences for the mechanism of lymphatic dissemination. Am J Pathol. 1997; 150:1951-7.
267. Clarijs R, Schalkwijk L, Ruiter DJ, de Waal RM. Lack of lymphangiogenesis despite coexpression of VEGF-C and its receptor Flt-4 in uveal melanoma. Invest Ophthalmol Vis Sci. 2001; 42:1422-8.
268. Mouta Carreira C, Nasser SM, di Tomaso E, Padera TP, Boucher Y, Tomarev SI, Jain RK. LYVE-1 is not restricted to the lymph vessels: expression in normal liver blood sinusoids and down-regulation in human liver cancer and cirrhosis. Cancer Res. 2001; 61:8079-84.
269. Papoutsi M, Siemeister G, Weindel K, Tomarev SI, Kurz H, Schachtele C, Martiny-Baron G, Christ B, Marme D, Wilting J. Active interaction of human A375 melanoma cells with the lymphatics in vivo. Histochem Cell Biol. 2000; 114:373-85.
270. Papoutsi M, Sleeman JP, Wilting J. Interaction of rat tumor cells with blood vessels and lymphatics of the avian chorioallantoic membrane. Microsc Res Tech. 2001; 55:100-7.
271. Schoppmann SF, Birner P, Studer P, Breiteneder-Geleff S. Lymphatic microvessel density and lymphovascular invasion assessed by anti-podoplanin immunostaining in human breast cancer. Anticancer Res. 2001; 21:2351-5.
272. Beasley NJ, Prevo R, Banerji S, Leek RD, Moore J, van Trappen P, Cox G, Harris AL, Jackson DG. Intratumoral lymphangiogenesis and lymph node metastasis in head and neck cancer. Cancer Res. 2002; 62:1315-20.
273. Jain RK, Fenton BT. Intratumoral lymphatic vessels: a case of mistaken identity or malfunction? J Natl Cancer Inst. 2002; 94:417-21.
274. Kurebayashi J, Otsuki T, Kunisue H, Mikami Y, Tanaka K, Yamamoto S, Sonoo H. Expression of vascular endothelial growth factor (VEGF) family members in breast cancer. Jpn J Cancer Res. 1999; 90:977-81.
275. Salven P, Lymboussaki A, Heikkila P, Jaaskela-Saari H, Enholm B, Aase K, von Euler G, Eriksson U, Alitalo K, Joensuu H. Vascular endothelial growth factors VEGF-B and VEGF-C are expressed in human tumors. Am J Pathol. 1998; 153:103-8.
276. Akagi K, Ikeda Y, Miyazaki M, Abe T, Kinoshita J, Maehara Y, Sugimachi K. Vascular endothelial growth factor-C (VEGF-C) expression in human colorectal cancer tissues. Br J Cancer. 2000; 83:887-91.
277. Andre T, Kotelevets L, Vaillant JC, Coudray AM, Weber L, Prevot S, Parc R, Gespach C, Chastre E. Vegf, Vegf-B, Vegf-C and their receptors KDR, FLT-1 and FLT-4 during the neoplastic progression of human colonic mucosa. Int J Cancer. 2000; 86:174-81.

278. Niki T, Iba S, Tokunou M, Yamada T, Matsuno Y, Hirohashi S. Expression of vascular endothelial growth factors A, B, C, and D and their relationships to lymph node status in lung adenocarcinoma. Clin Cancer Res. 2000; 6:2431-9.
279. Ohta Y, Nozawa H, Tanaka Y, Oda M, Watanabe Y. Increased vascular endothelial growth factor and vascular endothelial growth factor-c and decreased nm23 expression associated with microdissemination in the lymph nodes in stage I non-small cell lung cancer. J Thorac Cardiovasc Surg. 2000; 119:804-13.
280. Bunone G, Vigneri P, Mariani L, Buto S, Collini P, Pilotti S, Pierotti MA, Bongarzone I. Expression of angiogenesis stimulators and inhibitors in human thyroid tumors and correlation with clinical pathological features. Am J Pathol. 1999; 155:1967-76.
281. Ohta Y, Shridhar V, Bright RK, Kalemkerian GP, Du W, Carbone M, Watanabe Y, Pass HI. VEGF and VEGF type C play an important role in angiogenesis and lymphangiogenesis in human malignant mesothelioma tumours. Br J Cancer. 1999; 81:54-61.
282. Fellmer PT, Sato K, Tanaka R, Okamoto T, Kato Y, Kobayashi M, Shibuya M, Obara T. Vascular endothelial growth factor-C gene expression in papillary and follicular thyroid carcinomas. Surgery. 1999; 126:1056-61; discussion 1061-2.
283. Shushanov S, Bronstein M, Adelaide J, Jussila L, Tchipysheva T, Jacquemier J, Stavrovskaya A, Birnbaum D, Karamysheva A. VEGFc and VEGFR3 expression in human thyroid pathologies. Int J Cancer. 2000; 86:47-52.
284. Yonemura Y, Endo Y, Fujita H, Fushida S, Ninomiya I, Bandou E, Taniguchi K, Miwa K, Ohoyama S, Sugiyama K, Sasaki T. Role of vascular endothelial growth factor C expression in the development of lymph node metastasis in gastric cancer. Clin Cancer Res. 1999; 5:1823-9.
285. Eggert A, Ikegaki N, Kwiatkowski J, Zhao H, Brodeur GM, Himelstein BP. High-level expression of angiogenic factors is associated with advanced tumor stage in human neuroblastomas. Clin Cancer Res. 2000; 6:1900-8.
286. Tsurusaki T, Kanda S, Sakai H, Kanetake H, Saito Y, Alitalo K, Koji T. Vascular endothelial growth factor-C expression in human prostatic carcinoma and its relationship to lymph node metastasis. Br J Cancer. 1999; 80:309-13.
287. Kubo H, Fujiwara T, Jussila L, Hashi H, Ogawa M, Shimizu K, Awane M, Sakai Y, Takabayashi A, Alitalo K, Yamaoka Y, Nishikawa SI. Involvement of vascular endothelial growth factor receptor-3 in maintenance of integrity of endothelial cell lining during tumor angiogenesis. Blood. 2000; 96:546-53.
288. Dadras SS, Paul T, Bertoncini J, Brown LF, Muzikansky A, Jackson DG, Ellwanger U, Garbe C, Mihm MC, Detmar M. Tumor lymphangiogenesis: a novel prognostic indicator for cutaneous melanoma metastasis and survival. Am J Pathol. 2003; 162:1951-60.

# INDEX

## E



## F



## G



## H



## I



## L



## M



## N



## P

## *Q*



## *S*



## *T*



## *U*



## *V*